Benjamin/Cummings

Student Workbook for A.D.A.M.®

to accompany

A.D.A.M. Comprehensive
A.D.A.M. Standard
A.D.A.M. Standard Student Edition

Rose Leigh Vines, Ph.D.
California State University, Sacramento

The Benjamin/Cummings Publishing Company, Inc.

Menlo Park, California • Reading, Massachusetts
New York • Don Mills, Ontario • Wokingham, U.K. • Amsterdam
Bonn • Paris • Milan • Madrid • Sydney • Singapore • Tokyo
Seoul • Taipei • Mexico City • San Juan, Puerto Rico

Executive Editor: Johanna Schmid
Sponsoring Editor: Lisa Moller
Assistant Editors: Thor Ekstrom, Thomas Viano
Production Editor: Larry Olsen
Copy Editor: Anna Reynolds Trabucco
Compositor: Jeffrey Sargent
Manufacturing Supervisor: Merry Free Osborn
Cover Designer: Yvo Riezebos

Manufactured in the United States of America. Published simultaneously in Canada.

ISBN 0-8053-2115-2

1 2 3 4 5 6 7 8 9 10—CRS— 99 98 97 96 95

The Benjamin/Cummings Publishing Company, Inc.
2725 Sand Hill Road
Menlo Park, CA 94025

PREFACE

Welcome to the *Student Workbook for A.D.A.M.*® This workbook is a valuable interactive tool designed to enhance your knowledge of anatomy and physiology. This workbook can be used with Windows and Macintosh versions of A.D.A.M. Comprehensive, A.D.A.M. Standard, and their Student Editions. The order of topics reflects the organization of topics in *Human Anatomy and Physiology*, Third Edition, by Elaine N. Marieb. However, this workbook can be used with any human anatomy and physiology textbook.

Any time spent exploring the A.D.A.M.® anatomy software can help enrich your understanding of anatomy, but working through these guided exercises will help you to understand the cohesiveness of body systems, to see the most important relationships between structures and body systems, and to tie this powerful software tool to your anatomy course.

CUSTOMER SUPPORT

If you have problems installing or running A.D.A.M. software products for Macintosh or for Windows, call A.D.A.M. Technical Support at (404) 953-ADAM (2326) from 9 AM to 6 PM EST, Monday through Friday.

LEARNING AIDS

This workbook includes a number of pedagogical aids to enhance its use as a learning tool.

Introduction

The Introduction is an introduction to A.D.A.M. Standard. It includes tutorial exercises that will familiarize you with A.D.A.M. Before you start using the workbook, take some time to work through the guided exercises in the Introduction in order to become accustomed to the software. The Introduction includes information on the Macintosh and Windows versions of A.D.A.M. Standard.

Objectives

Each chapter includes Student Objectives. These serve as a guide to let you know what topics will be covered and what points you wiyou have completed a chapter.

Overview

Each chapter begins with exercises that introduce you to the body system you are studying. For these exercises, be sure to follow the instructions that guide you to the system overview in A.D.A.M. Standard.

Exercises

Each chapter includes a set of comprehensive questions for that particular body system. There are a variety of question types, including multiple choice, true or false, fill-in the blank, matching, and questions that require you to sketch anatomical structures.

Icons

You will notice that some questions include a pencil icon or a human figure icon. Questions that include the pencil icon ask you to sketch a figure in order to answer the question. Questions that include the human figure icon approach anatomy regionally and ask you to synthesize information that you have learned from lecture and from your anatomy text.

Answers

Answers for all exercises (except for sketch exercises) are included at the back of the workbook.

ACKNOWLEDGMENTS

I would like to thank Elaine Marieb for her review and helpful suggestions for the *Student Workbook for A.D.A.M.* I would also like to thank reviewers who gave direction and suggestions for questions in the workbook: Connie L. Allen, Edison Community College; Clare Hays, Metropolitan Sate College of Denver; William Hightower, Southside Virginia Community College; and Carlene Tonini, College of San Mateo. I am very grateful to Sue Marek, technical writer, who provided clear instructions in the Introduction of the workbook on how to use A.D.A.M. Standard.

I would also like to thank the staff at A.D.A.M. Software, Inc., for their helpful suggestions and guidance. My sincere thanks goes to the editorial staff at Benjamin/Cummings for their support: Johanna Schmid, Executive Editor; Lisa Moller, Senior Acquisitions Editor; Robin Heyden, Executive Producer; and Thor Ekstrom, Assistant Editor. I am especially grateful to my editor, Thomas Viano, who gave me constant encouragement and guidance throughout this project. His cheerful attitude and sound advice were greatly appreciated. I also thank Larry Olsen, Senior Production Editor, for his contributions to the workbook.

THE A.D.A.M. SCHOLAR SERIES®

- ***A.D.A.M. Essentials***
  Specially designed for high school and introductory courses, self-tutorial, and review, this CD-ROM presents a wealth of anatomical drawings superimposed upon each other. Containing approximately 200 layers in Anterior and Posterior Views, A.D.A.M. Essentials uses limited scientific terminology, making it a useful resource in the school library, learning resource center, or laboratory.

- ***A.D.A.M. Standard***
  A.D.A.M. Standard is an exciting interactive CD-ROM program developed for teaching human anatomy to undergraduate students in allied health, nursing, and pre-med programs. It incorporates over 400 layers in Anterior, Posterior, Medial, and Lateral views as well as system views. Terminology is consistent with the undergraduate course. A student version of A.D.A.M. Standard is also available, and it can be used independently and as a reference. The student version comes with a complimentary copy of the *Student Workbook for A.D.A.M.*

- ***A.D.A.M. Comprehensive***
  Originally designed for use in medical schools and teaching hospitals, A.D.A.M. Comprehensive is a rich resource with over 2,000 layers that allows students and instructors to explore the human body at the highest level of detail.

- *A.D.A.M.–Benjamin/Cummings Interactive Physiology*
  Two new CD-ROMS—one on the cardiovascular system and one on the muscular system—are designed to clarify the difficult topics of physiology. These two CD-ROMS put the power of multimedia—animation, sound, narration, and video—to its best use in explaining the physiological concepts and processes students find most difficult to understand.

A.D.A.M. software and texts by Elaine Marieb—the perfect combination for learning anatomy and physiology.

- ***Human Anatomy & Physiology,*** **Third Edition**
  To enhance your study of anatomy and physiology, selected end-of-chapter questions are highlighted with the "walking man" icon and can be answered by consulting A.D.A.M. Standard.

- ***Study Guide To Accompany Human Anatomy and Physiology,*** **Third Edition**
  To create a powerful learning program, selected figures in the Study Guide are highlighted with the A.D.A.M. "walking man" icon and are keyed to an appendix that guides you to the relevant views and layers in A.D.A.M. Standard.

- ***Human Anatomy and Physiology Lab Manual,*** **Fourth Edition**
  (also available in Cat and Fetal Pig Versions)
  A new A.D.A.M. appendix correlates some of the anatomical lab observations with corresponding sections in the A.D.A.M. anatomy software.

- ***Anatomy and Physiology Coloring Workbook***

- ***Essentials of Anatomy and Physiology,*** **Fourth Edition**

- ***Human Anatomy***
  Elaine N. Marieb and Jon Mallatt
  A special insert links selected figures and review questions to A.D.A.M. Standard by guiding you to the relevant view and layer in the software.

  If you have questions about purchasing texts by Elaine Marieb, call The Benjamin/Cummings Publishing Company, Inc., at 800-950-2665. If you have questions about purchasing A.D.A.M. Comprehensive, A.D.A.M. Standard or A.D.A.M.--Benjamin/Cummings Interactive products, product shipping and distribution, and training, call The Benjamin/ Cummings Publishing Company, Inc., at 800-950-2665 (ask for A.D.A.M. support).

CONTENTS

USING THIS WORKBOOK

Welcome to the *Student Workbook for A.D.A.M.* This workbook is appropriate for use with the software programs A.D.A.M. Comprehensive, A.D.A.M. Standard, and the Student Editions of these products. This workbook is for students in the health and life sciences curriculum, including students of allied health, nursing, or premedical health sciences.

Directions for using A.D.A.M. software products refer to A.D.A.M. Standard. However, this guide is compatible with A.D.A.M. Comprehensive as well as the Student Editions of A.D.A.M. Comprehensive and A.D.A.M. Standard. This workbook is designed to help you use A.D.A.M. Standard to learn about human anatomy. It is divided into two parts. The first part provides an introduction to A.D.A.M. Standard. The second part contains exercises for learning anatomy using A.D.A.M. Standard.

INTRODUCTION

A.D.A.M. Standard is a multimedia program for learning about human anatomy—A.D.A.M. stands for Animated Dissection of Anatomy for Medicine. This program contains hundreds of layers of precise anatomical illustrations. These are combined with an interactive toolbox you use to examine and identify hundreds of structures in the human body. Because of its flexibility, there are many ways you can use A.D.A.M. Standard to study anatomy. You can:

- Explore the full human anatomy, layer by layer, from four different views and by gender.
- Explore the 12 major systems of human anatomy.
- Read text overviews of each system that explain their processes and functions.
- Point and click to identify different structures.
- View cross sections, magnetic resonance images (MRIs), and histology slides.
- Hear the pronunciation of different terms.
- Simulate operating room procedures.

GETTING STARTED

The exercises and descriptions in this workbook are based on the assumption that A.D.A.M. Standard is already installed and running properly on your computer. If necessary, refer to the *User's Guide* that accompanied your software package for more information on:

- Operating system requirements
- Installation
- Locking and Unlocking Fig Leaves
- Customer support and technical support

WHAT YOU NEED TO KNOW

Before using A.D.A.M. Standard, you need to be familiar with basic operations in the computer environment you are running. A.D.A.M. Standard is available for both the Macintosh® and Windows™ systems. Basic operations for these systems include:

- Starting the computer and operating system.
- Launching applications.
- Using desktop icons along with the Finder in Macintosh, or the File Manager and DOS in Windows, to manage documents, applications, and disks.
- Clicking, holding, dragging, and scrolling operations using a mouse or track ball.
- Displaying pop-up and pull-down menus and selecting menu commands.
- Opening, closing, scrolling, moving, selecting, and resizing windows.
- Using the clipboard.
- Using a CD-ROM drive.

For more information on these tasks, refer to the manual that came with your computer and operating system.

WORKING WITH MACINTOSH OR WINDOWS

A.D.A.M. Standard for the Macintosh and A.D.A.M. Standard for Windows are very similar programs and have almost identical user interfaces. There are, however, some differences. The information in this workbook describes A.D.A.M. Standard features as they appear and function on the Macintosh. Differences with the Windows program are noted when applicable.

At the end of this Introduction is a section about differences between the primary windows and on-line help features in Windows and Macintosh.

LEARNING TO USE A.D.A.M.

This section contains four brief exercises to help you begin using A.D.A.M. Standard. These are:

- Exercise 1: Exploring the Primary Window and the Full Anatomy
- Exercise 2: Studying Anatomical Systems
- Exercise 3: Finding a Structure
- Exercise 4: Exploring the Library
- Exercise 5: Exploring the Operating Room

These exercises will demonstrate useful features and tools for studying anatomy using A.D.A.M. Standard. As you are doing these exercises, you may find it useful to refer to the Quick Reference information that appears on pages 28–35.

EXERCISE 1: EXPLORING THE PRIMARY WINDOW AND THE FULL ANATOMY

This exercise is designed to acquaint you with the major features of the Primary window and the Full Anatomy view. You begin by opening the A.D.A.M. Standard software.

Starting the A.D.A.M. Standard Session in Macintosh

Double–click the icon labeled "A.D.A.M. Standard" in the A.D.A.M. folder on the hard disk (shown below). If you purchased A.D.A.M. Standard Student Edition, double click on the icon labeled "A.D.A.M. Standard Student Edition."

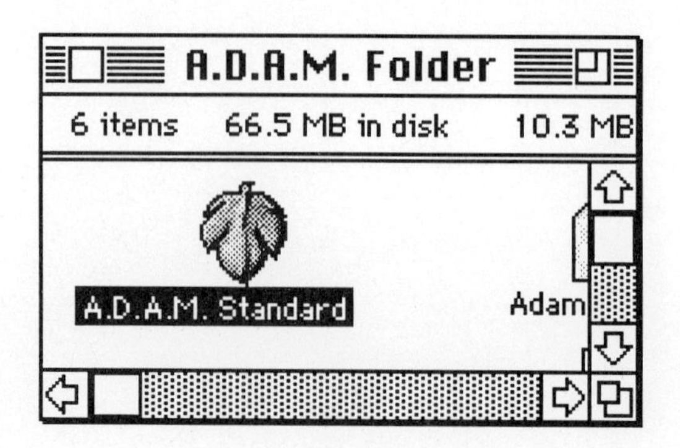

The A.D.A.M. Standard startup screen appears, followed by the A.D.A.M. Standard screen. The A.D.A.M. Standard screen consists of three primary areas: the Primary window, the Library window, and the Structure List window. As you progress through the workbook, you will be using various tools in each window. The tools in each window are described in the following exercises.

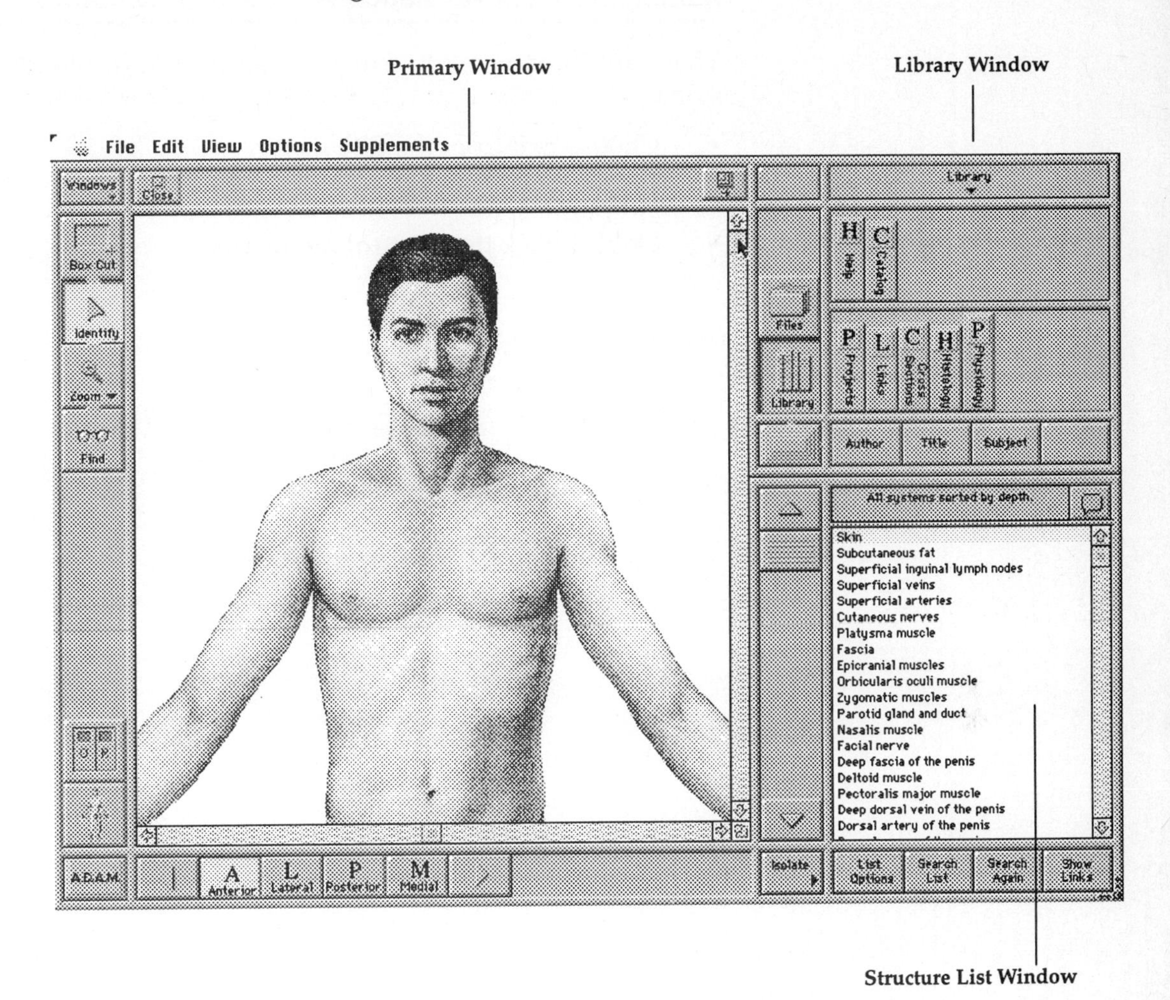

Quitting the A.D.A.M. Standard Session in Macintosh

You can quit A.D.A.M. Standard software at any time by selecting Quit from the File menu.

Starting the A.D.A.M. Standard Session in Windows

Double-click the icon labeled "A.D.A.M. Standard" in the A.D.A.M. Standard group. The A.D.A.M. Standard introduction screen appears, followed by the A.D.A.M. Standard screen.

Exiting the A.D.A.M. Standard Session in Windows

There are three ways to exit from an A.D.A.M. Standard session.

- Choose Exit from the **File** menu.
- Press **Alt+F4**.
- Double-click the **Control menu box**.

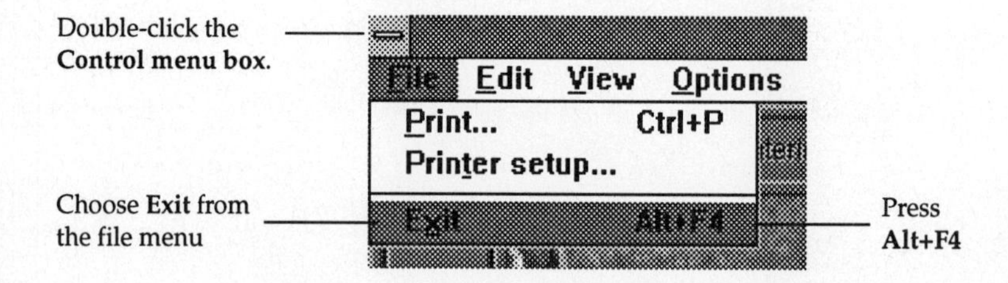

Expanding the Window to Fit Your Monitor

If the A.D.A.M Standard screen is not expanded to the full size of your monitor screen, resize it by dragging the red-colored size box in the lower right corner to the location you want. *(This feature is available in Macintosh only.)*

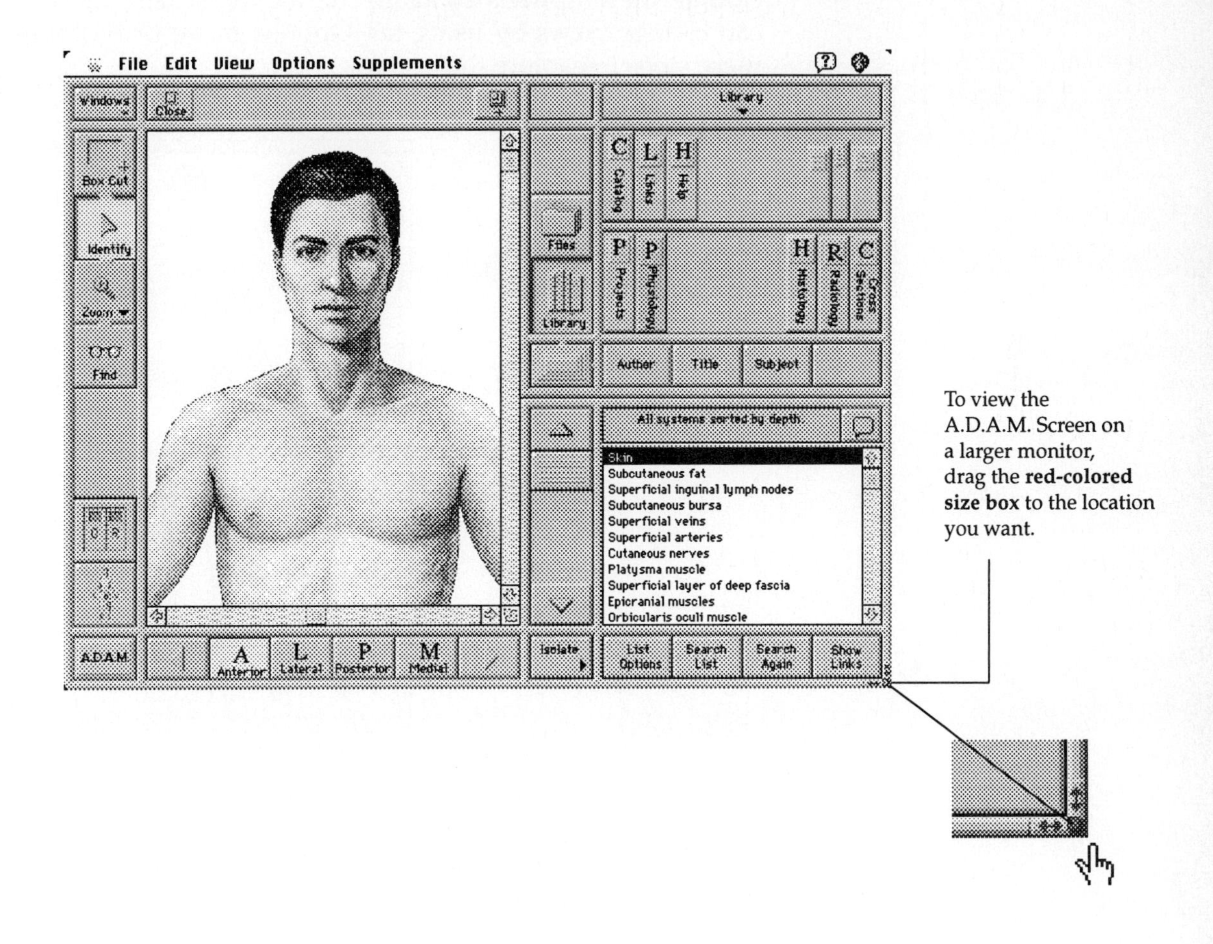

To view the A.D.A.M. Screen on a larger monitor, drag the **red-colored size box** to the location you want.

Setting Options for the Anatomical Image

After you open A.D.A.M. Standard, an image of the full human anatomy is displayed in the Primary window. You can customize several aspects of its appearance, including gender, skin tone, fig leaves, and anatomical view. Windows users must change these options by using the Options menu. Mac users can change views by using the Options menu or by following these steps to explore the options.

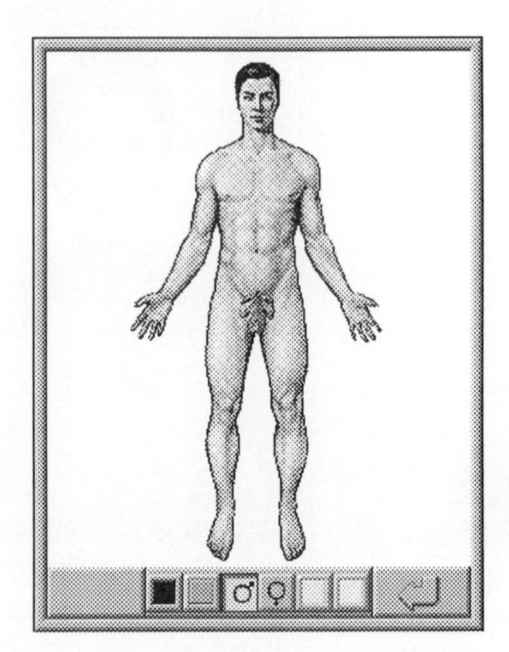

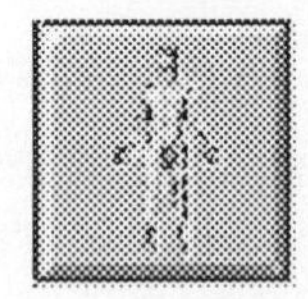

1. Click the **Adam Man** button to display the **Preference Window**.
2. If the **Fig Leaves** option is not locked during installation, you can add or remove the fig leaves by clicking on them.
3. Then, click the different skin tones and the two gender symbols to change the appearance of the figure.
4. Click the **Return Button** to set the options and return to the Primary window. The **Return Arrow** is located in the lower right corner of the window.
5. At the bottom of the Primary window are buttons for selecting the anatomical view (**Anterior**, **Lateral**, **Posterior**, and **Medial**). Experiment with the different views, and then set the view to **Anterior** by clicking the **A** button.

Using the Structure List

The Structure List window lists the structures available for examination in the Full Anatomy view. When you start A.D.A.M. Standard, structures are listed according to depth, beginning with the skin.

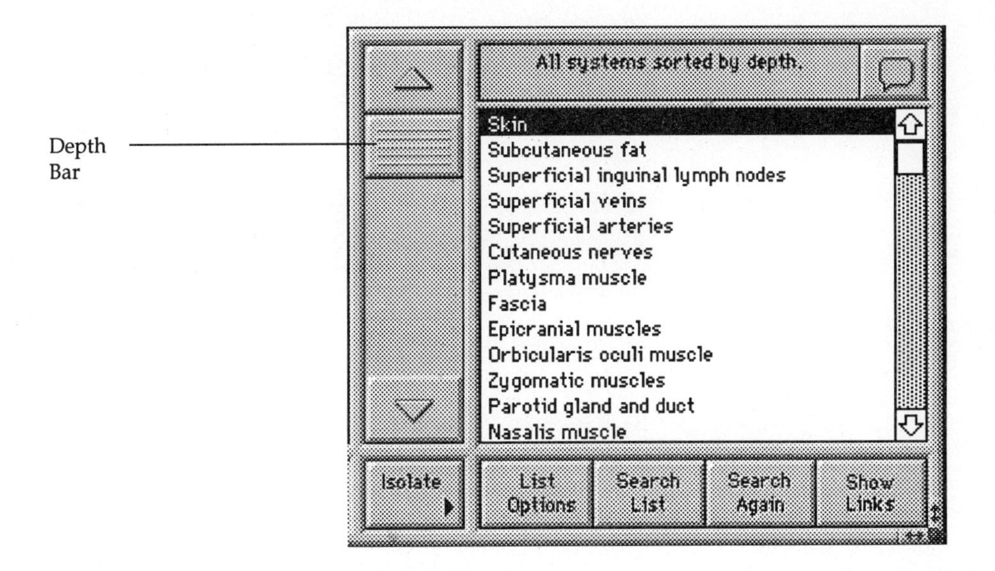

Follow these steps to experiment with the Structure List and **Isolate** button.

1. Begin at the Skin layer. (If it is not already displayed in the Image Area, select it by clicking on Skin in the Structure List.)
2. Drag the **Depth Bar** up and down to different depths, stopping at different positions. See how both the list and the image change. Return to the Skin layer.
3. Click different items in the Structure List to change the depth (you can scroll the list to display more structures), then return again to the Skin layer.

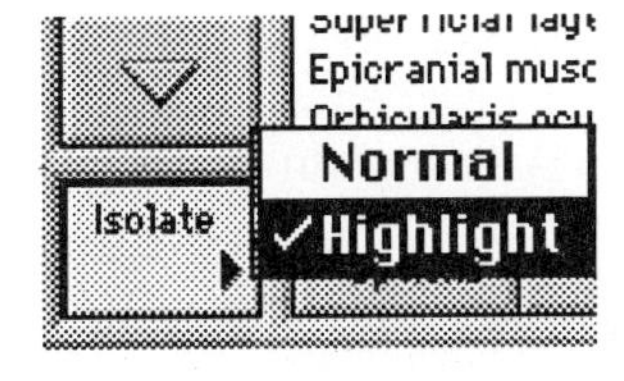

4. Next, in the Structure List, click Zygomatic muscles. Now, click and hold the **Isolate** button to display a pop-up menu and select **Highlight**. This action will "gray out" all structures except for Zygomatic muscles, the one you have selected in the list.
5. Click other structures in the list or move the **Depth Bar**. Each structure that you select is shown in color to isolate it for identification. Then, click and hold the **Isolate** button again to display the pop-up menu and select **Normal**. The image is returned to full color.
6. Return to the Skin layer.

Examining and Identifying Structures

One of the best ways to use A.D.A.M. Standard is to explore and identify parts of the anatomy. Follow these steps to learn about the **Identify, Box Cut,** and **Zoom** tools for examining the anatomy.

Identify Tool

1. Click and hold the Status Bar at the top of the Primary window. If you are using Windows, just click on the status bar. This displays an alphabetized structure list.

2. Scroll to the top of the list and select Abdominal aorta and branches to display that structure in the Image Area.
3. Select the **Identify** tool at the left by clicking it.
4. Point to a structure on the anatomical image, and click on it. This displays that structure's name in the Status Bar at the top of the Primary window.
5. Next, click and hold the mouse button on a structure, for example, the right kidney. This displays the **Identify** pop-up menu. The name of a specific renal structure is shown in bold type, under the system to which it belongs.
6. Release the button, move to another part of the kidney, and click and hold the mouse button to display the **Identify** pop-up menu again.
7. This time, when the pop-up menu appears, drag the pointer over the lower arrow at the right to reveal a sub menu. Drag down to select Best View. This displays the best view of a structure in the current anatomical view.

   If you choose *Best View for the System Anatomy*, the Image Area highlights the system and dims the rest of the anatomy. The Text Overview window replaces the Structure List window. To return to the Full Anatomy, click the **Return Arrow** on the Primary window panel.

   If you choose *Best View for the structure*, the Image Area shows the structure, and the Structure List highlights the name.
8. Click and hold on another structure. Drag down to select **Highlight**.

   This accents a selected structure and shows it in relation to other structures by dimming them. (To undo the highlight, press the **Isolate** button and select **Normal** from the pop-up menu.)
9. Click and hold on another structure. Drag down to select Histology.

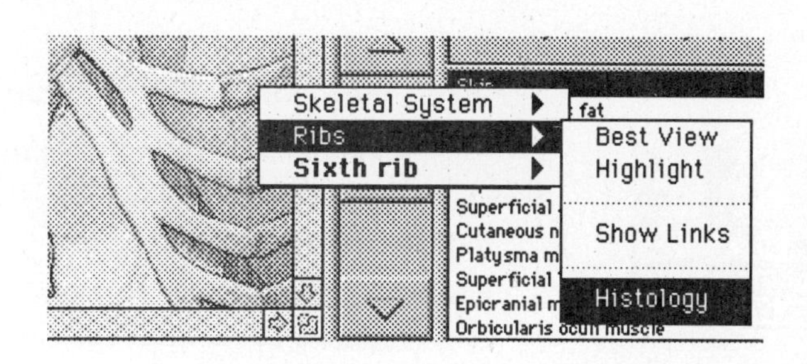

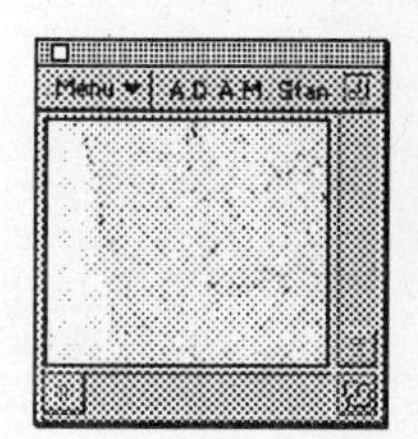

A window appears that displays the histology of the selected structure. When you are finished viewing the histology, close it by clicking the white Close box in the upper left corner of the window.

10. Return to the Skin layer by dragging the **Depth Bar** to the top.

Now, continue the exercise to learn about the **Box Cut** tool.

Box Cut Tool

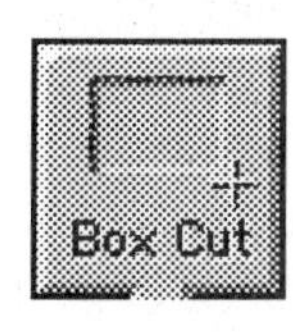

1. Select the **Box Cut** tool by clicking it.
2. If the abdominal region is not visible in the Image Area, use the scroll bars to bring it into the window.
3. Click and drag the cross-bar to draw a rectangle across the abdominal region of the anatomical image.
4. This reveals the next layer below the Skin, i.e., the Subcutaneous fat, within the box cut.
5. Next drag the **Depth Bar** down to different levels to reveal other structures at other depths.

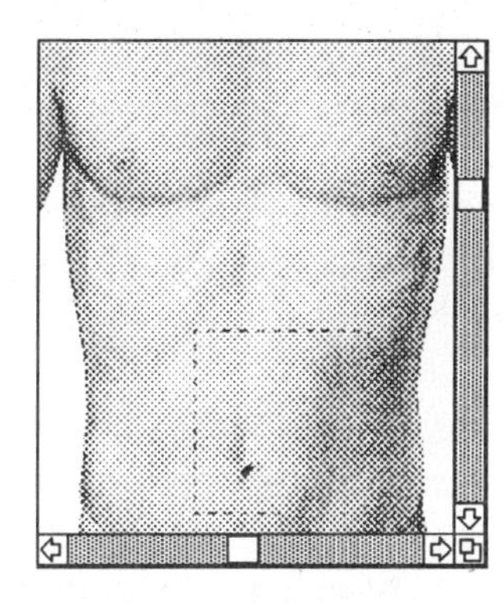

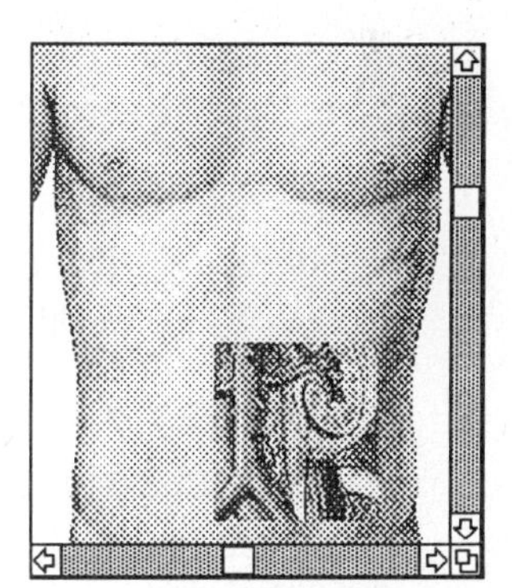

6. When you find a structure that you are interested in exploring, click the **Identify** tool and then click on the structure to display its name in the Status Bar.
7. Then, click and hold the pointer on the structure to display the **Identify** pop-up menu. From the structure sub menu, select **Highlight**. This highlights that structure. To return to normal, click and hold the **Isolate** button in the Structure List window and select **Normal**.

Zoom Tool

Now, use the **Zoom** tool to examine the box cut area.

1. Click and hold the **Zoom** tool to display its pop-up menu.
2. Select **Magnifying Glass**, and the pointer changes to a magnifying glass.
3. Click the pointer over the image area and release the mouse button. Windows users must click and hold over the image area. The pointer changes to a 2-inch-diameter lens enclosing a plus sign (+). The lens magnifies the area beneath it as it moves. Click again to close the lens, and then click again to redisplay the lens.
4. Display the **Zoom** pop-up menu again. This time select **Zoom Out**. The pointer changes to a small magnifying glass with a minus sign (-) in the center.
5. Click on the area of the anatomy you want to become the center of the view. The image is reduced.
6. Display the **Zoom** pop-up menu again. This time select **Zoom In**. The pointer changes to a magnifying glass with a plus sign (+) in the center.
7. Click on the area of the anatomy you want to become the center of the view. The image is enlarged.
8. Continue practicing with the **Identify** tool, the **Box Cut** tool, and the **Zoom** tool as you move through different layers of the anatomical image.
9. When you are finished practicing, click the **Identify** tool to change the cursor back to a pointer. Next, return to the Skin layer by selecting it in the Structure List.

EXERCISE 2: STUDYING ANATOMICAL SYSTEMS

In the previous exercise, you have been studying and moving through the Full Anatomy view. Now we shall focus on studying specific anatomical systems.

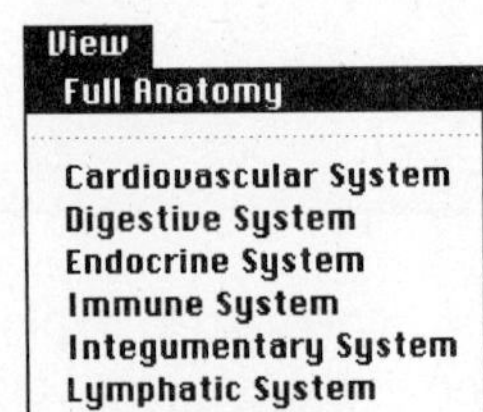

Viewing a System Anatomy

1. Click and hold the **View** menu to display its contents. Notice that it lists the Full Anatomy you are now viewing and 12 System Anatomy views.
2. Select Cardiovascular System. Notice that the Image Area changes, as do the available tools.
3. The Structure List window has been replaced by the Text Overview window. Scroll through the Text Overview to read about the cardiovascular system.
4. Click the **Search Text** button to display a dialog box for locating a word in the text overview. Click the **Search Again** button to find the next occurrence of a searched-for word.

Search Text

5. Click the **Labels** button at the lower left corner of the Text Overview window. This displays labels for the different structures in the system.
6. Select the Identify tool. Click and hold the pointer over any highlighted structure in the system to display the **Identify** pop-up menu. In the System view, the **Identify** menu shows the name of the structure and the pronunciation option. Drag down to the **Pronounce** command to hear the pronunciation of the structure's name.
7. To view more of the image, click the **Expand** button at the upper right of the Primary window. The image will expand to the size of your screen.

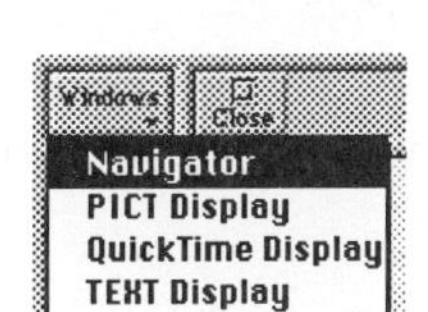

Using the Navigator

An easy way to move around the image is to use the **Navigator**.

1. Click and hold the **Windows** button at the upper left of the Primary window to display a pop-up menu. The **Windows** button is only available in Macintosh. If you are using Windows, click the Navigator button.
2. Select **Navigator** from the menu to display the Navigator window. In this window a box defines the portion of the anatomical image that is visible on your screen.
3. Click and drag the box to another area of the image to display it. When you are finished practicing, close the **Navigator** by clicking the white Close box in the upper left corner. If you are using Windows, double click the white Close box.
4. Click the **Expand** button again to restore all windows to their original state. If the head and torso are not visible, use the scroll bars to bring them into view.

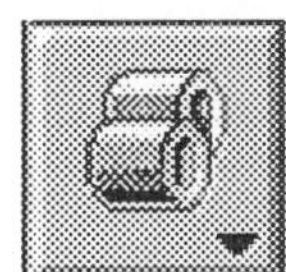

The System view also provides options that are specific to different systems. These can be selected using the **System-Specific Options** button at the lower left of the Primary window or by using the Options menu. In Cardiovascular System, for example, you select options for viewing arteries and/or veins by clicking and holding on the button. Select one of the options.

Practice using the tools you learned in this exercise while viewing other anatomical systems. When you are finished exploring the systems, return to the Full Anatomy view by selecting Full Anatomy from the View menu. Return to the Skin layer by clicking Skin in the Structure List.

EXERCISE 3: FINDING A STRUCTURE

When using A.D.A.M. Standard, most of your tasks will involve finding structures within the human anatomy. There are several tools for finding those structures. Once you have located a structure, you can use the **Highlight**, **Identify**, and **Zoom** tools to examine it further.

Using the Alphabetical Structure List

As mentioned earlier, one simple way to find a structure is to use the Alphabetical Structure List in the Full Anatomy view. For example, to find the pancreas:

1. Click and hold the Status Bar of the Primary window to display the Alphabetic Structure List. If you are using Windows, just click on the status bar.

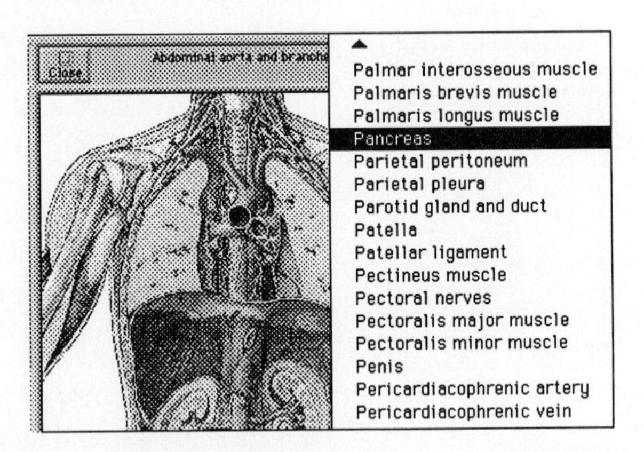

2. Scroll up or down the list until Pancreas is highlighted. Release the mouse button to select Pancreas. The anatomical image changes to the best view of that structure.

Searching in the Structure List

Another way to search is to use the Structure List in the Full Anatomy view. You can scroll and select items in the list. The anatomical image changes to the best view of that structure. To use the Search feature:

1. Click the **Search List** button.
2. In the dialog box that appears, enter a name, for example, *aorta*, in the **Search For** text field.

Search For:

aorta

Cancel OK

3. Click **OK**. A.D.A.M. Standard highlights the first occurrence of that word in the Structure List.
4. Click the **Search Again** button, and A.D.A.M. Standard highlights the next occurrence of that word.
5. When you have found the specific aortic structure you wish to examine, click it in the list. The anatomical image changes to the best view of that structure. If you are using Windows, the structure will automatically appear.

Searching in the Text Overview

If you are in the System Anatomy view, you can search for information about specific structures in the Text Overview. In this example, you find more information about ventricles.

1. Select Cardiovascular System from the View menu.
2. In the Text Overview window, click the **Search Text** button.
3. In the dialog box that appears, enter the word *Ventricle* in the **Search For** text field.
4. Click **OK**, and A.D.A.M. Standard highlights the first occurrence of that word in the Text Overview of that system.
5. Click the **Search Again** button, and A.D.A.M. Standard highlights the next occurrence of that word.

Using the Find Tool

The **Find** tool in the Full Anatomy view is a good way to identify and find several components of a structure at once. This example shows how to find all available views of the lungs.

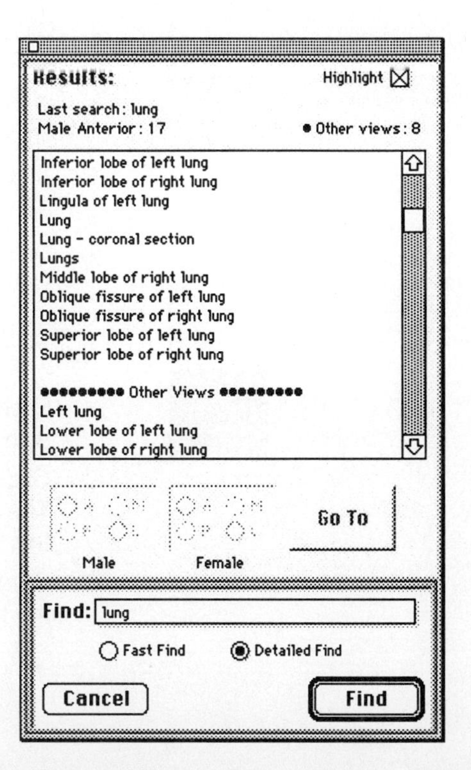

1. Click the **Find** tool to display the Find dialog box.
2. Type *Lung* in the text field, click the **Detailed Find** button, and then click **Find**.
3. When the Results box appears, you can scroll its list to search for all structure views and system text overviews associated with that word.
4. Click the **Highlight** checkbox in the upper right corner to highlight the structure you select within the full anatomy.
5. Select an item in the list and click **Go To** to display it in the Image Area. The Results list remains open until you close it. Click the Close box in the upper left corner of the dialog box to close it.

Now you can practice using the Find features. When you are finished, return to the Full Anatomy view, at the Skin layer.

EXERCISE 4: EXPLORING THE LIBRARY

The Library window contains documents that support your work with A.D.A.M. Standard as well as additional books. The exercise below provides a tour of library resources.

Viewing Cross Sections and MRIs

In addition to the Full and System Anatomy views, A.D.A.M. Standard provides cross-section illustrations and magnetic resonance images (MRIs) that extend the length and width of the human anatomy. To view these images:

1. Click on the Cross Section book in the Library window to open it. The Primary window displays the Full Cross Section image. If you are using Windows, double click to open this book.

2. Click on an Eye icon to display the Cross Section window for its associated region.
3. In the Cross Section window, you can click and hold the pointer on a region to identify it.
4. Next, use the **Zoom/Magnifying Glass** tool to get a closer look. *(Not available in Windows.)*
5. When you are finished viewing the cross section, close its window by clicking the white Close box in the upper left corner.

You can also change from cross-section illustrations to magnetic resonance images.

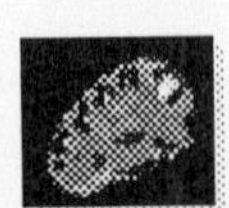

1. Click the **MRI** button in the Full Cross Section window. If you are using Windows, double click the button.
2. Click on an Eye icon to display an MRI window for that region. The **Identify** pointer and **Zoom/Magnifying Glass** tool function in this window as well. *(Not available in Windows.)*
3. When you are finished viewing an MRI, close its window by clicking the white Close box in the upper left corner.

Viewing Histologies

The Library includes an extensive collection of histology slides. As described in Exercise 1, you can view a single histology slide for a specific structure using the **Identify** pop-up menu. If you are using a Macintosh, another way to use the histology feature is to display histology slides in pairs and compare them with each other and with the Full Anatomy views. This method is described below.

For example, to compare the microscopic structures of the internal organs:

1. Go to the Full Anatomy view and display the Gallbladder structure by selecting it in the Alphabetic Structure List.
2. Then, click the Histology book in the Library window to open it. A window appears with two histology slides. *(Note: The Histology book is not available in Windows. Windows users see p. 27 for more information on viewing histologies.)*
3. Click the **Radio** button (◉) in the upper histology slide to select it. (The **Radio** button is in the lower left corner of the slide.)

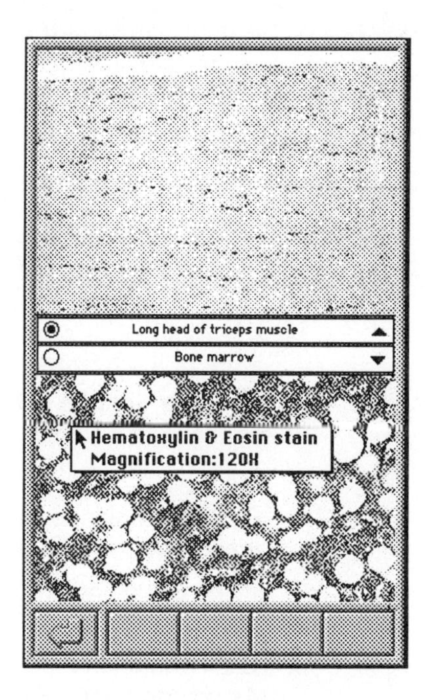

4. Using the **Identify** tool, click and hold the pointer on a structure in the anatomy image to display the **Identify** pop-up menu.
5. Select **Histology** from the sub menu for that structure. The histology for that structure appears in the upper histology slide.
6. Next, click the **Radio** button in the lower histology window to select it.
7. Use the **Identify** tool and pop-up menus to select the histology for a different structure in the anatomy image so that it appears in the lower histology slide.

   You can also choose a histology slide by system or alphabetically by clicking and holding the pointer on an arrow indicator.

This way you can compare the two histologies with each other and with their associated structures. When you are finished viewing histologies, click the **Return Arrow** at the bottom left corner of the Histology window.

EXERCISE 5: EXPLORING THE OPERATING ROOM

When you are in the Full Anatomy view, you can click the **OR** button to open the Operating Room panel. The Image Area remains, but you now have a different set of tools to work with. Use Operating Room tools to simulate surgical procedure steps.

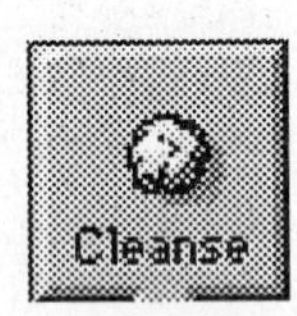

Cleanse

Click this tool and the pointer changes to a sterilizing swab. This option returns you to the Skin layer. Click and drag to swab the skin in preparation for surgery.

Syringe

Click this tool and the pointer changes to a syringe. Click on a structure to mimic a local anesthetic infusion. (*Not available in Windows.*)

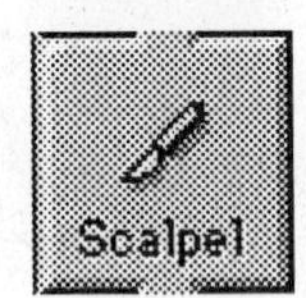

Scalpel

Click this tool and the pointer changes to a surgical blade for incisions.

Using the Scalpel Tool

1. Locate the area you want to cut using the **Navigator** tool, scroll bars, or the **Expand** button.
2. Click the **Scalpel** tool on the Operating Room panel. The pointer changes to a scalpel with a narrow cut width.

   –or–

   Click and hold the **Scalpel** tool. A pop–up menu appears; choose the cut width. *(Not available in Windows.)*
3. In the Structure List, click the visible structure you want to cut through or to.
4. Click one end of the cut site, and drag a line to mark the incision. As you drag the tool, the incision line appears.
5. Release the mouse button. For Macintosh users, the cut appears as specified, and the Structure List highlights the name. For Windows users, the pointer moves to the center of the incision. Move the pointer to define the retraction width and click on the diamond shape to make the cut appear.

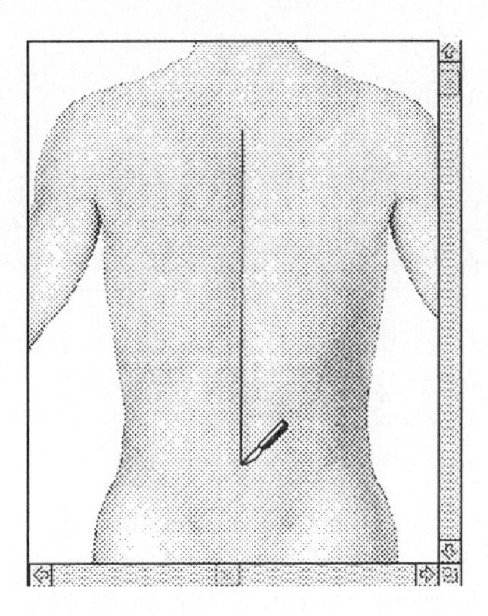

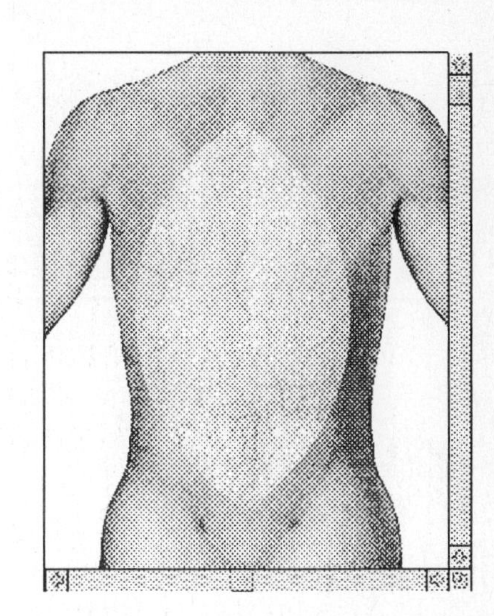

Increasing or Decreasing the Cut Depth

There are three ways to change the depth of the cut:

- Move the **Depth Bar** up or down.
- Select another structure from the Structure List.
- Create another scalpel cut within the first.

To undo all cuts and return to the Skin, drag the **Depth Bar** to the top.

Laser

Click this tool and the pointer changes to a laser. Then, click and drag to simulate laser surgery.

Cauterize

Click this tool to select it. Then, click and drag to simulate delivery of electric current for burning away skin or other tissues. *(Not available in Windows.)*

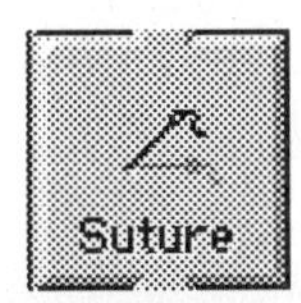

Suture

Click this tool to select it. Then, click and drag to simulate sewing.

Identify

Click this button and the pointer changes to the **Identify** tool.

Return Arrow

Click this button to redisplay the regular Primary window tools.

MACINTOSH REFERENCE: ONLINE HELP

ONLINE HELP

In addition to this workbook, you can refer to multimedia online Help from within A.D.A.M. Standard at any time. Help provides demonstrations on navigating around A.D.A.M. Standard, using buttons and tools, and creating books and links. *(Windows users see p. 23 for information about online Help. Student Edition users refer to your User's Guide.)*

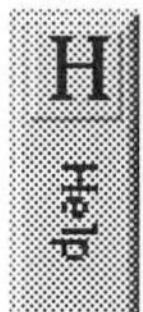

To start Help, click the Help book on the Library bookshelf. The Help dialog box appears.

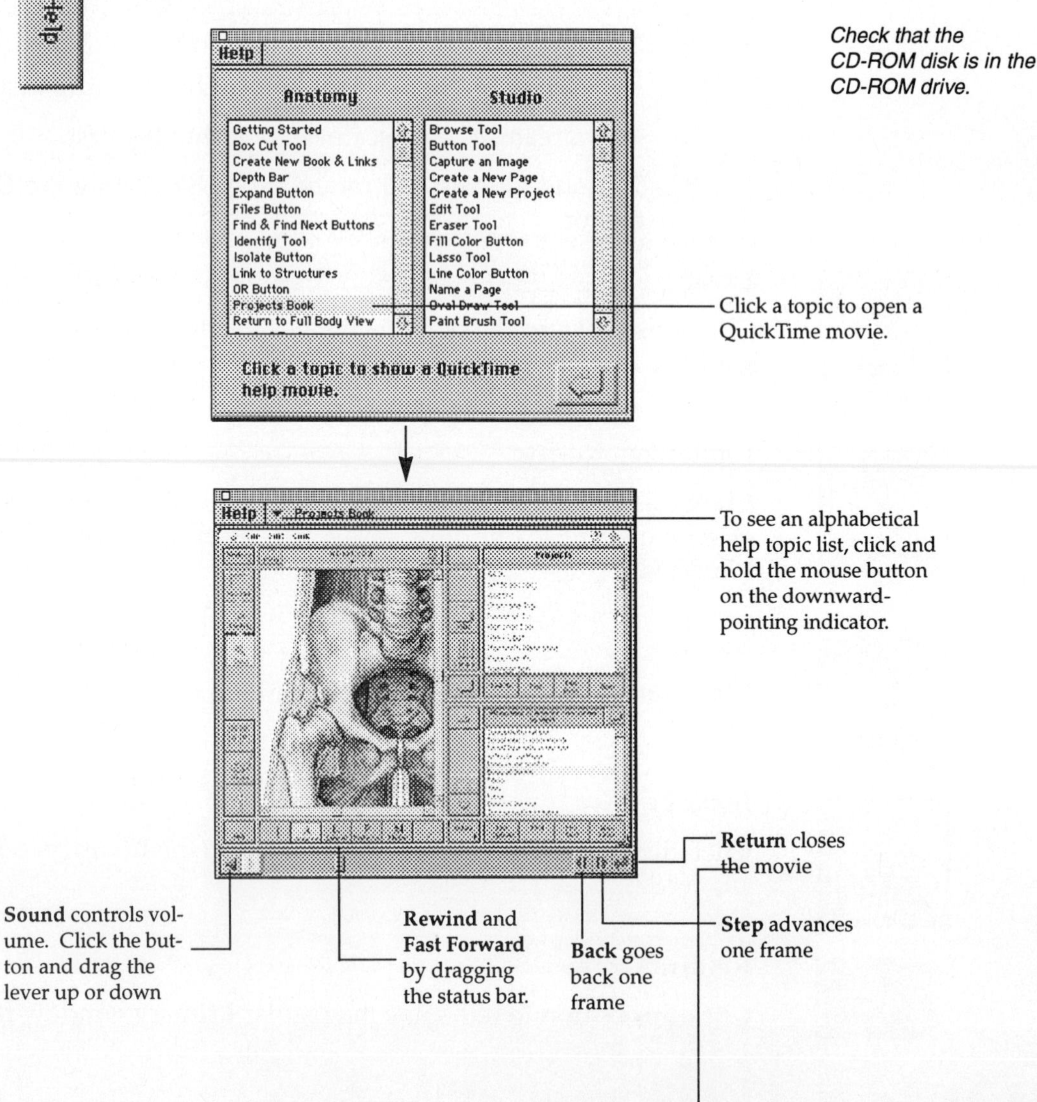

BALLOON HELP

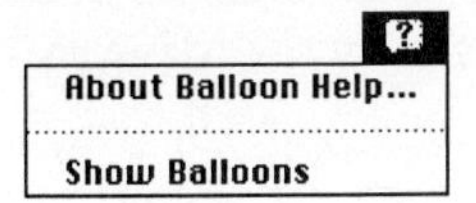

Refer to Balloon Help for brief explanations of icons, buttons, tools, and other A.D.A.M. Standard features. To start this context-sensitive Help system, choose Show Balloons from the **Help** menu at the upper right corner of the A.D.A.M. Standard screen. Placing the pointer over an object displays information about it. Balloon Help also explains Macintosh commands and functions while A.D.A.M. Standard is running.

PRIMARY WINDOW

The Primary window is the focal point for all your activities. The tools and menu commands you choose affect the image you see. This section describes the tools, menu commands, and other features in the Primary window.

The Image Area

The main feature of the Primary window is the Image Area. This area shows either the Full Anatomy view (male or female) containing all dissectable structures or one of 12 System Anatomy views, which are nondissectable.

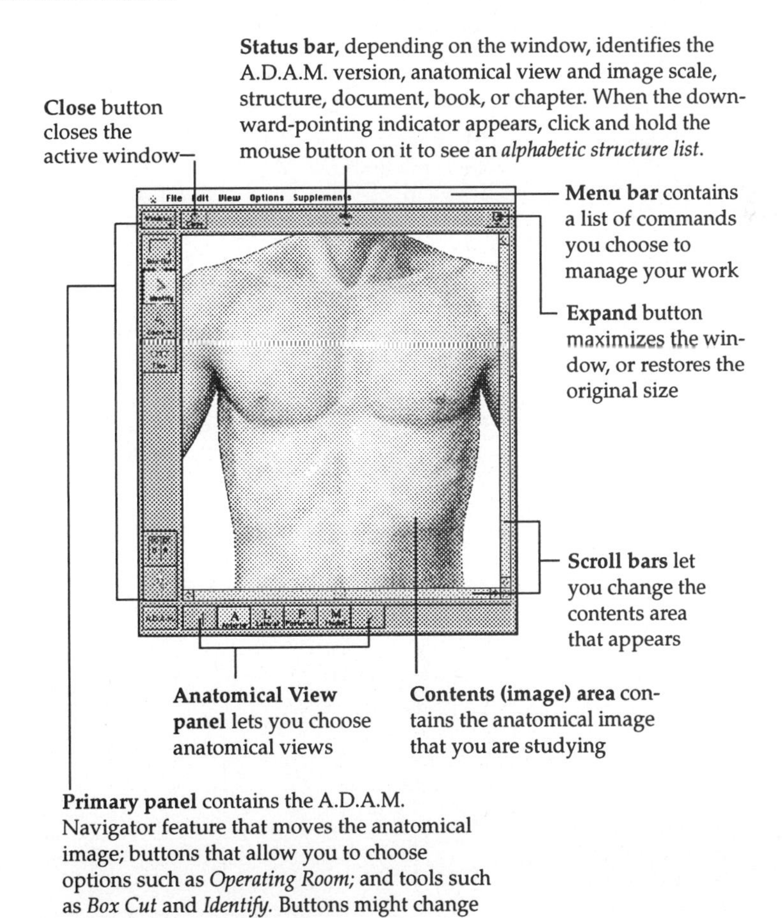

Primary Window Menu Commands

You use A.D.A.M. Standard menus to choose options and execute commands. These menus and their functions are shown in the diagram below.

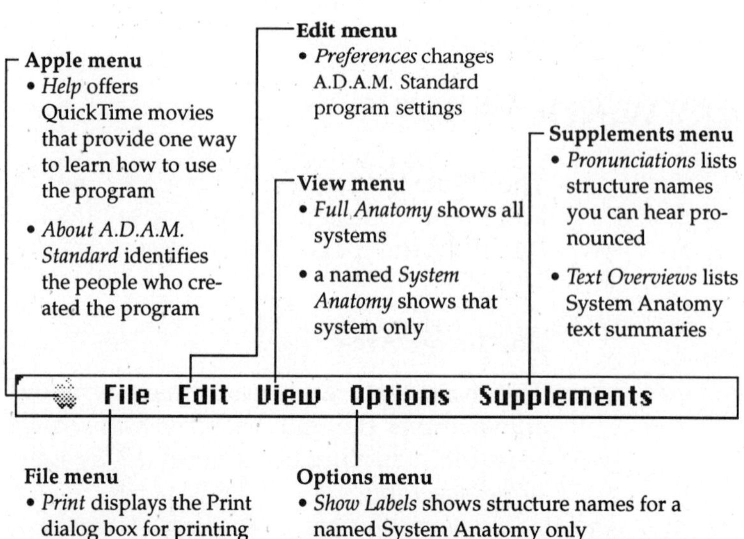

File menu

- *Print* displays the Print dialog box for printing the A.D.A.M. Screen, including books, anatomical images, and links
- *Quit* ends the A.D.A.M. Standard session

Options menu

- *Show Labels* shows structure names for a named System Anatomy only
- *Anterior (front), Posterior (back), Lateral (side), and Medial (midsagittal)* show the selected view
- *Female* and *Male* show the selected gender
- *Skin Tone* shows four skin tones and corresponding facial features
- *Fig Leaves*, a discretion option, lets users cover genitals and female breasts with fig leaves. If the option is dimmed and checked, the option was locked during installation and you cannot remove the fig leaves.
- *Arteries and Veins, Arteries Only*, or *Veins Only* show these structures for the Cardiovascular System Anatomy only
- *With Ribs* shows these structures for the Skeletal System Anatomy front view only; *Without Ribs* removes these structures

ONLINE HELP IN WINDOWS

You can refer to online Help from within A.D.A.M. Standard at any time. Help includes step-by-step procedures and explains icons, buttons, tools, and other A.D.A.M. Standard features. Using Help, you can display topics and navigate through them, jump to related Help topics, see definitions of terms, and search for and print information. To start Help, choose among Help menu items.

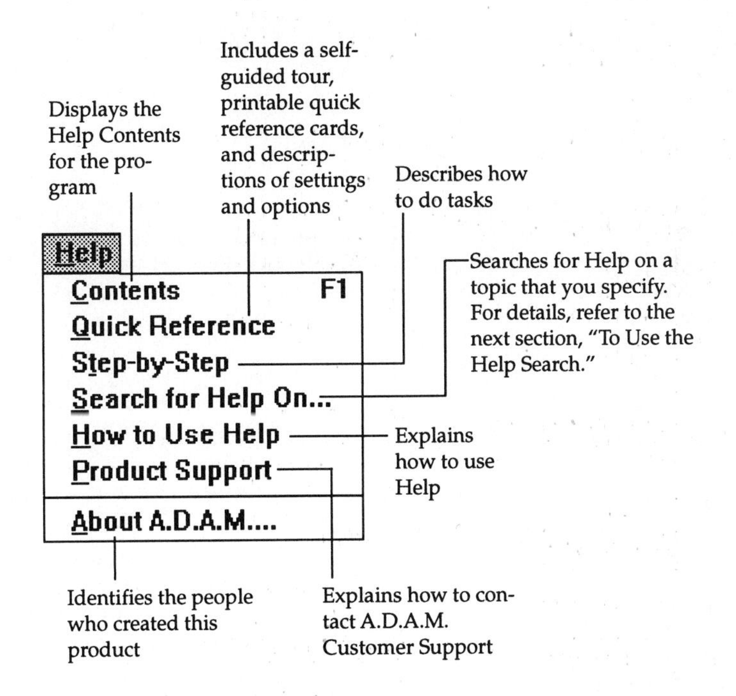

For example, chosing Contents from the **Help** menu displays a Help window. Command buttons are located below the menu bar in the window.

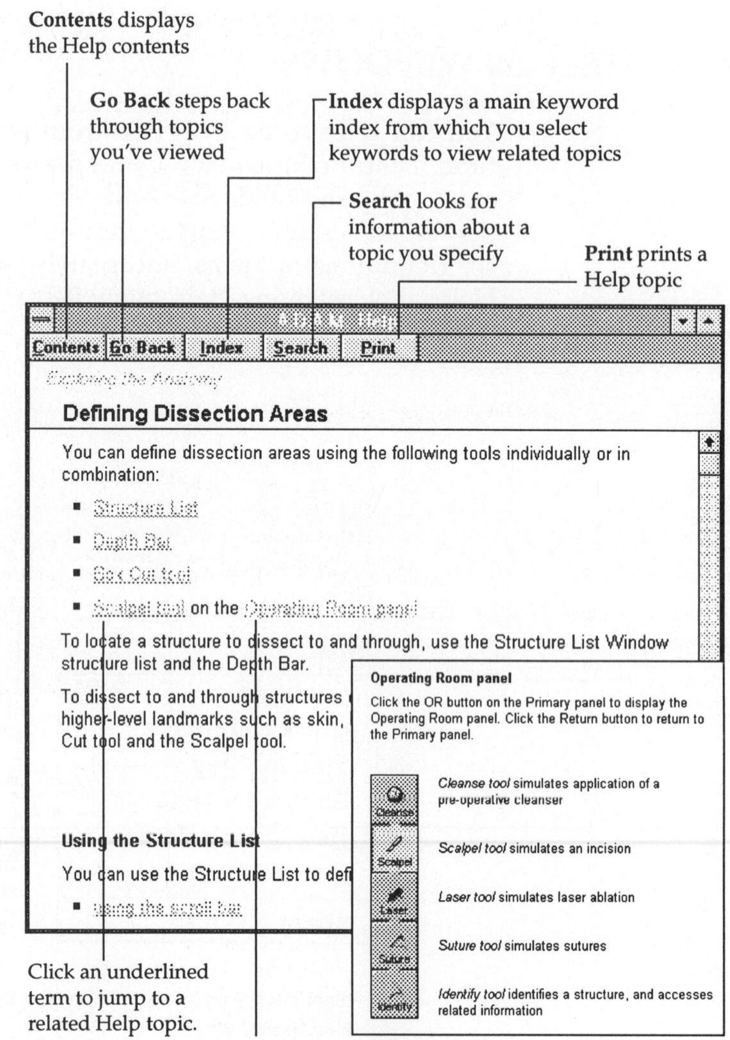

Using the Help Search Feature

Follow the three steps below to learn how to use the Help Search feature.

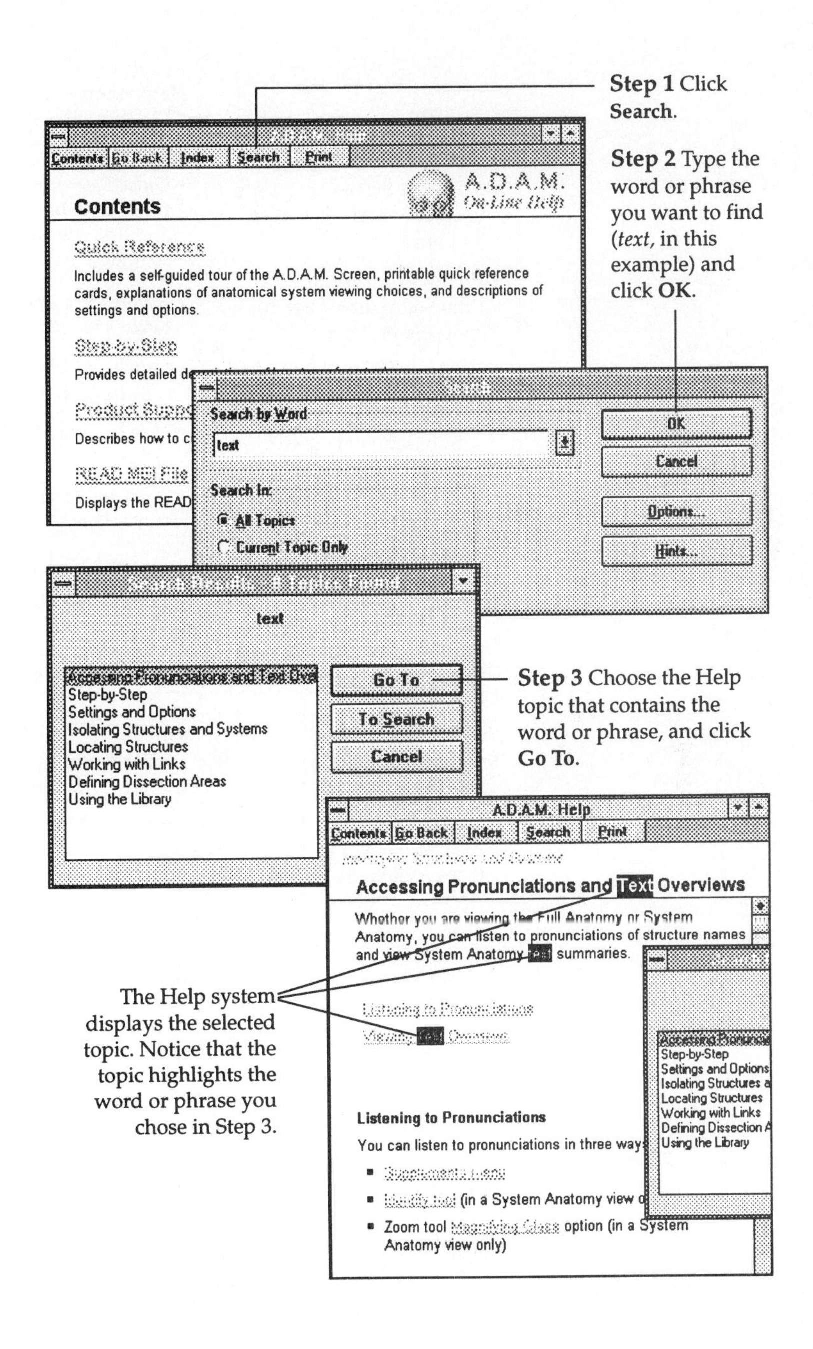

A.D.A.M. STANDARD MENUS IN WINDOWS

File menu
- *Print* displays the Print dialog box for printing the A.D.A.M. Screen, including books, anatomical images, and links
- *Printer setup* displays the Print Setup dialog box for choosing preferences
- *Exit* ends the A.D.A.M. Standard session

Supplements menu
- *Pronunciations* lists structure names you can hear pronounced
- *Text Overviews* lists System Anatomy text summaries

Help menu
- *Contents* displays the Help Contents for the program
- *Quick Reference* includes a self-guided tour, printable quick reference cards, and descriptions of settings and options
- *Step-by-Step* describes how to do tasks
- *Search for Help* on searches for Help on a topic that you specify
- *How to Use Help* explains how to use Help
- *Product Support* explains how to contact A.D.A.M. Customer Support
- *About A.D.A.M.* identifies the people who created this product

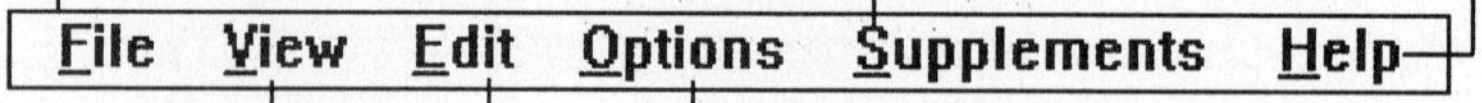

File View Edit Options Supplements Help

View menu
- *Full Anatomy* shows all systems
- a named *System Anatomy* shows that system only

Edit menu
- *Preferences* changes A.D.A.M. Standard program settings

Options menu
- *Show Labels* shows structure names for a named System Anatomy only
- *Anterior (front), Posterior (back), Lateral (side), and Medial (midsagittal)* show the selected view
- *Female* and *Male* show the selected gender
- *Skin Tone* shows four skin tones and corresponding facial features
- *Fig Leaves*, a discretion option, lets users cover genitals and female breasts with fig leaves. If the option is dimmed and checked, the option was locked during installation and you cannot remove the fig leaves.
- *Arteries and Veins, Arteries Only,* or *Veins Only* show these structures for the Cardiovascular System only
- *With Ribs* shows these structures for the Skeletal System front view only; *Without Ribs* removes these structures

VIEWING HISTOLOGIES IN WINDOWS

The Windows version of A.D.A.M. provides access to histology selections in the Full Anatomy. From the Full Anatomy Identify pop-up menu, choose the structure, then choose Histology from the submenu, and release the mouse button.

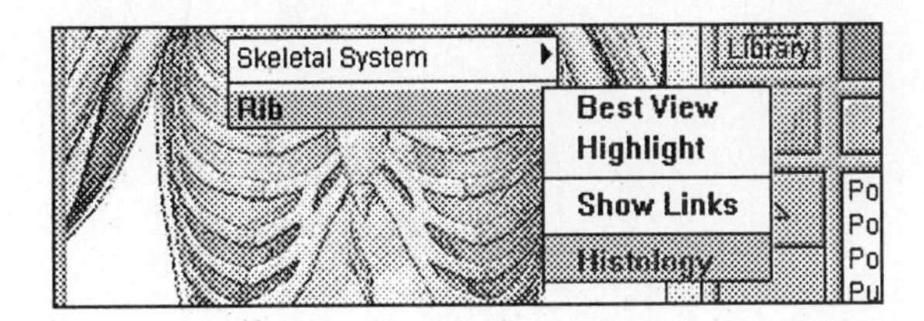

The histology slide appears in a floating window, and the mouse pointer changes to a magnifying glass when placed on the image.

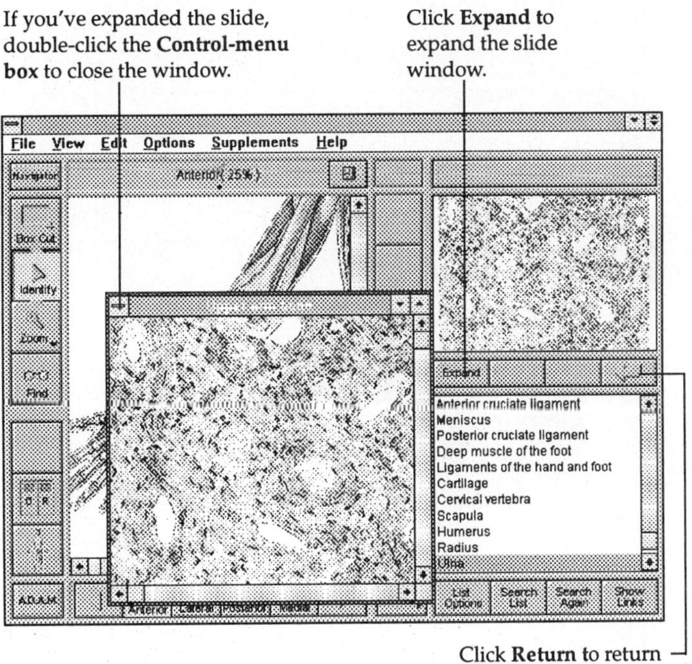

A.D.A.M. Comprehensive for Windows™

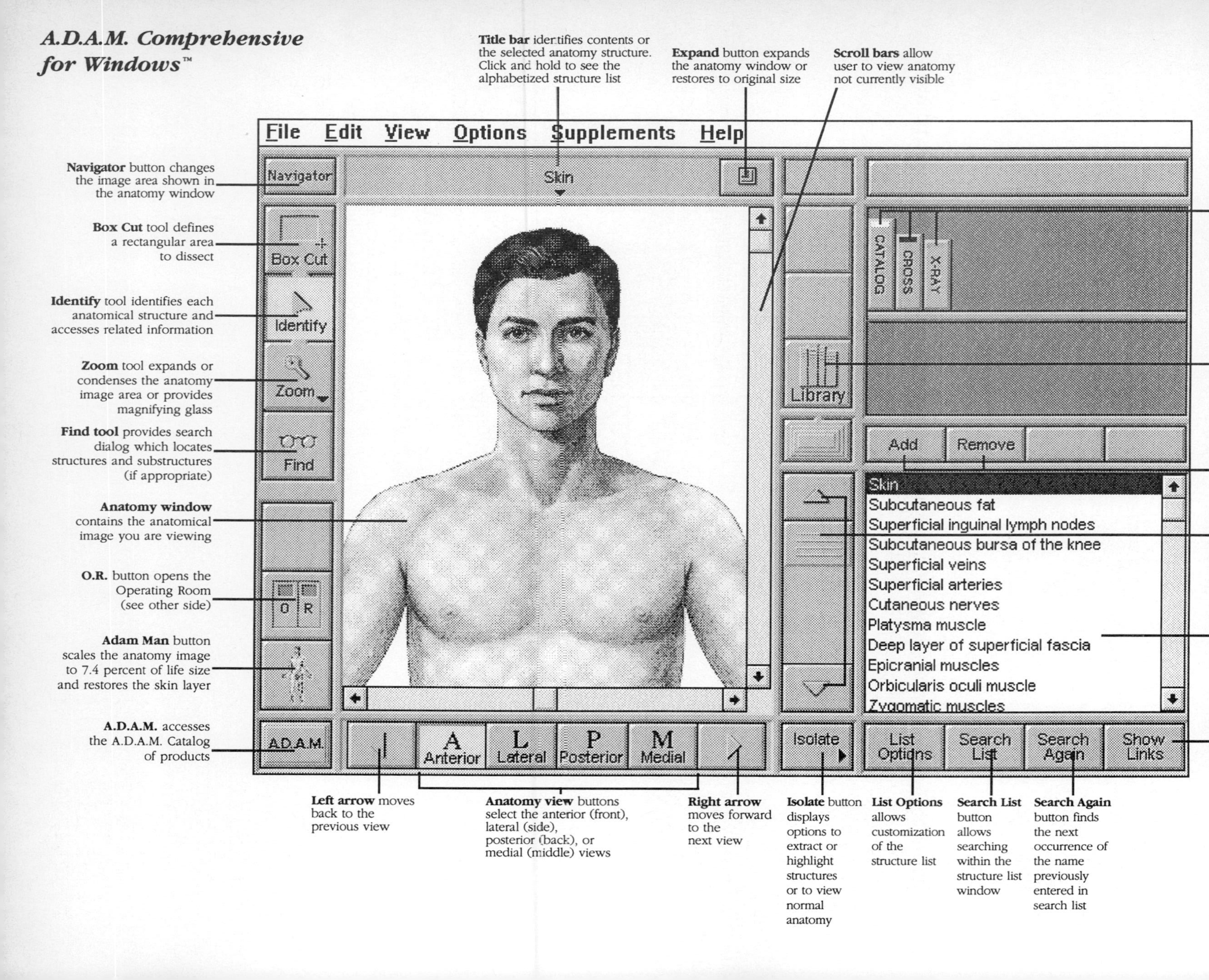

Options pull-down menu allows one to choose preferences

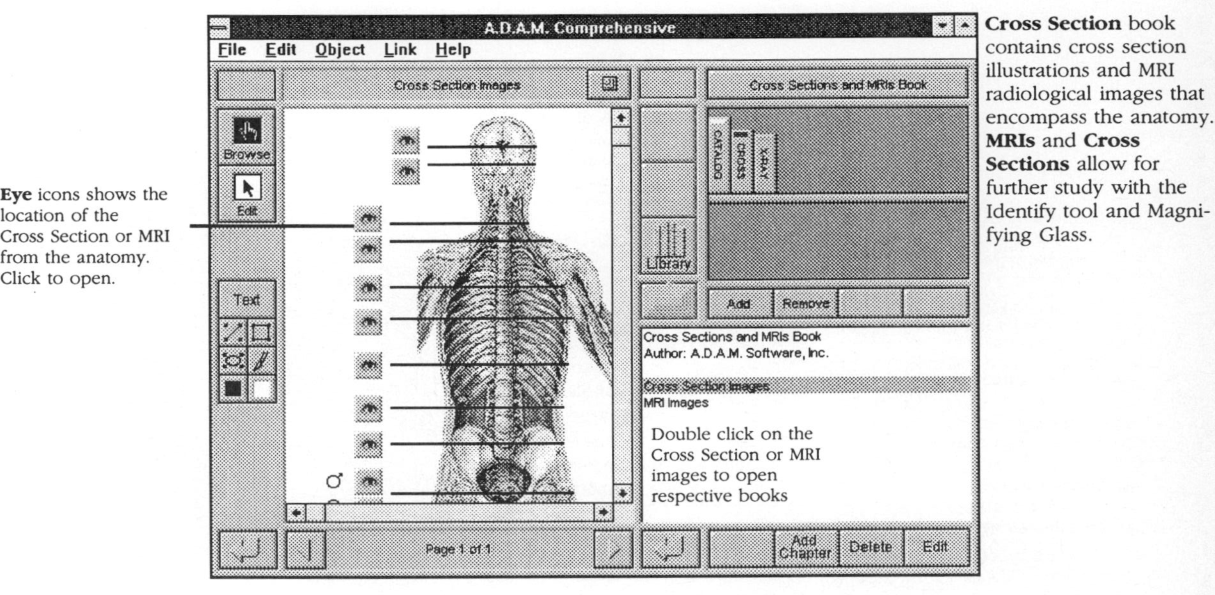

Select the anterior (front), lateral (side), posterior (back), or medial (middle) views, or medial or lateral arm

Choose male or female anatomy

Select among four skin tones and corresponding facial features

Fig leaves option removes or adds the fig leaves from the reproductive anatomy

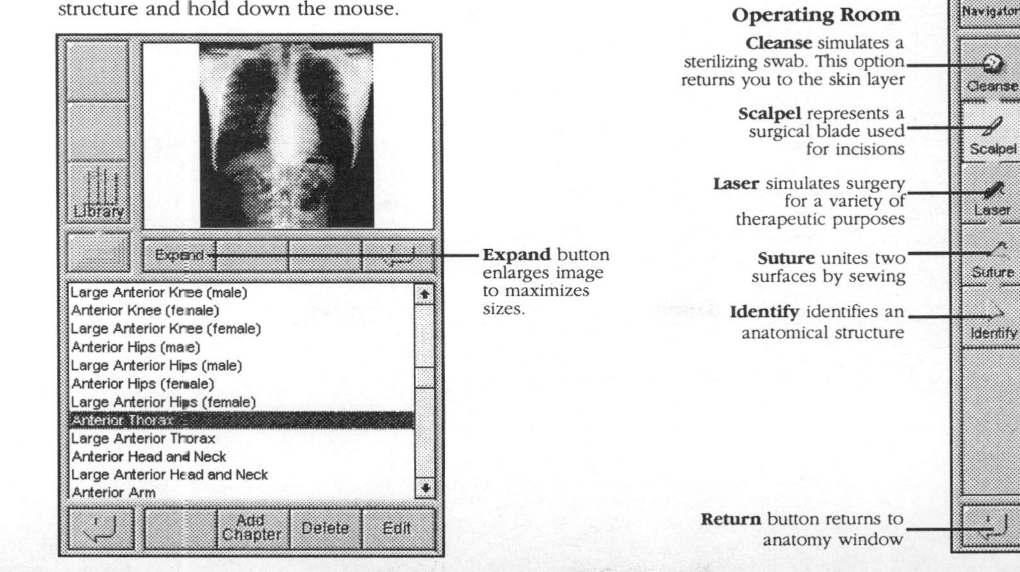

Eye icons shows the location of the Cross Section or MRI from the anatomy. Click to open.

Double click on the Cross Section or MRI images to open respective books

Cross Section book contains cross section illustrations and MRI radiological images that encompass the anatomy. **MRIs** and **Cross Sections** allow for further study with the Identify tool and Magnifying Glass.

Histology slides show the relationship between histologies and gross anatomy. Histology can be accessed using the Identify tool in the anatomy window.

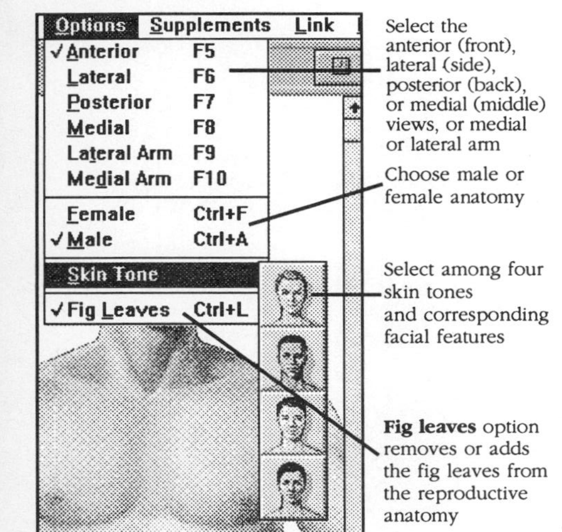

X-Ray book contains Radiologies of the entire body. Identify tool identifies structures in the radiology image. Place the tool on a structure and hold down the mouse.

Expand button enlarges image to maximizes sizes.

Operating Room

Cleanse simulates a sterilizing swab. This option returns you to the skin layer

Scalpel represents a surgical blade used for incisions

Laser simulates surgery for a variety of therapeutic purposes

Suture unites two surfaces by sewing

Identify identifies an anatomical structure

Return button returns to anatomy window

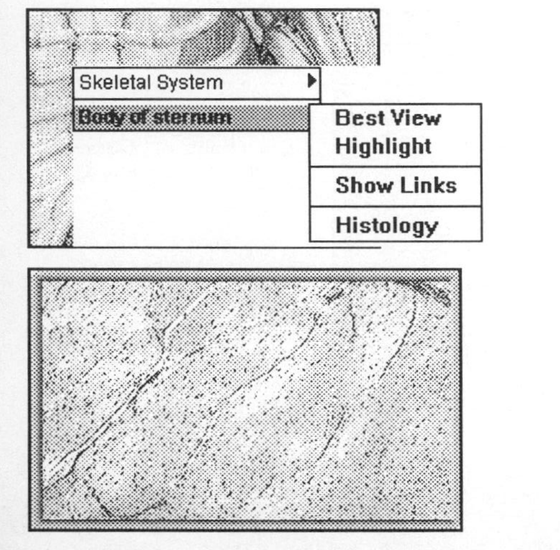

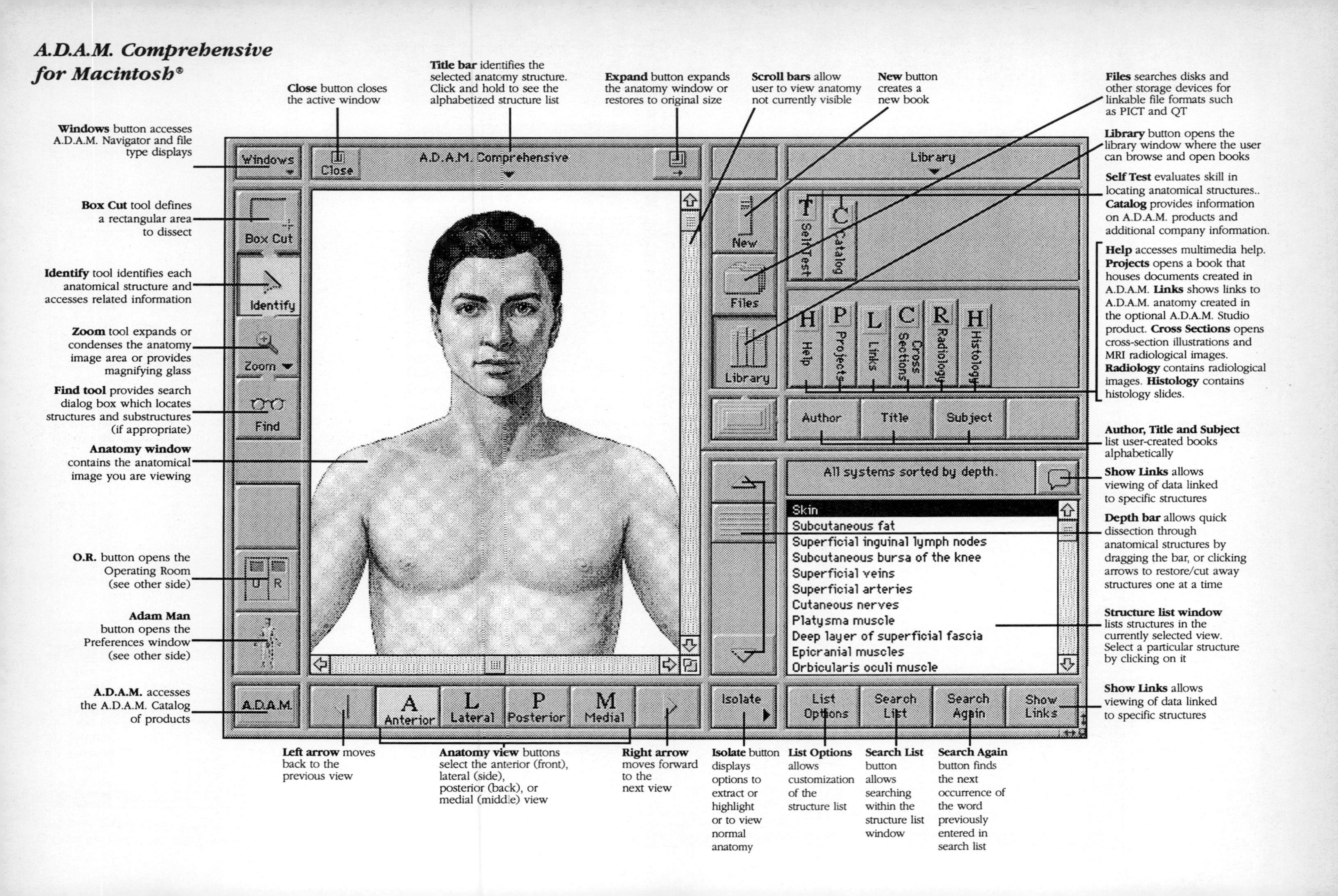
A.D.A.M. Comprehensive
for Macintosh®
Windows button accesses A.D.A.M. Navigator and file type displays
Box Cut tool defines a rectangular area to dissect
Identify tool identifies each anatomical structure and accesses related information
Zoom tool expands or condenses the anatomy image area or provides magnifying glass
Find tool provides search dialog box which locates structures and substructures (if appropriate)
Anatomy window contains the anatomical image you are viewing
O.R. button opens the Operating Room (see other side)
Adam Man button opens the Preferences window (see other side)
A.D.A.M. accesses the A.D.A.M. Catalog of products
Close button closes the active window
Title bar identifies the selected anatomy structure. Click and hold to see the alphabetized structure list
Expand button expands the anatomy window or restores to original size
Scroll bars allow user to view anatomy not currently visible
New button creates a new book
Files searches disks and other storage devices for linkable file formats such as PICT and QT
Library button opens the library window where the user can browse and open books
Self Test evaluates skill in locating anatomical structures.. Catalog provides information on A.D.A.M. products and additional company information.
Help accesses multimedia help. Projects opens a book that houses documents created in A.D.A.M. Links shows links to A.D.A.M. anatomy created in the optional A.D.A.M. Studio product. Cross Sections opens cross-section illustrations and MRI radiological images. Radiology contains radiological images. Histology contains histology slides.
Author, Title and Subject list user-created books alphabetically
Show Links allows viewing of data linked to specific structures
Depth bar allows quick dissection through anatomical structures by dragging the bar, or clicking arrows to restore/cut away structures one at a time
Structure list window lists structures in the currently selected view. Select a particular structure by clicking on it
Show Links allows viewing of data linked to specific structures
Left arrow moves back to the previous view
Anatomy view buttons select the anterior (front), lateral (side), posterior (back), or medial (middle) view
Right arrow moves forward to the next view
Isolate button displays options to extract or highlight or to view normal anatomy
List Options allows customization of the structure list
Search List button allows searching within the structure list window
Search Again button finds the next occurrence of the word previously entered in search list
Windows
Close
A.D.A.M. Comprehensive
Library
Box Cut
Identify
Zoom
Find
New
Files
Library
Self Test
Catalog
Help
Projects
Links
Cross Sections
Radiology
Histology
Author
Title
Subject
All systems sorted by depth.
Skin
Subcutaneous fat
Superficial inguinal lymph nodes
Subcutaneous bursa of the knee
Superficial veins
Superficial arteries
Cutaneous nerves
Platysma muscle
Deep layer of superficial fascia
Epicranial muscles
Orbicularis oculi muscle
A.D.A.M.
Anterior
Lateral
Posterior
Medial
Isolate
List Options
Search List
Search Again
Show Links

Preferences window

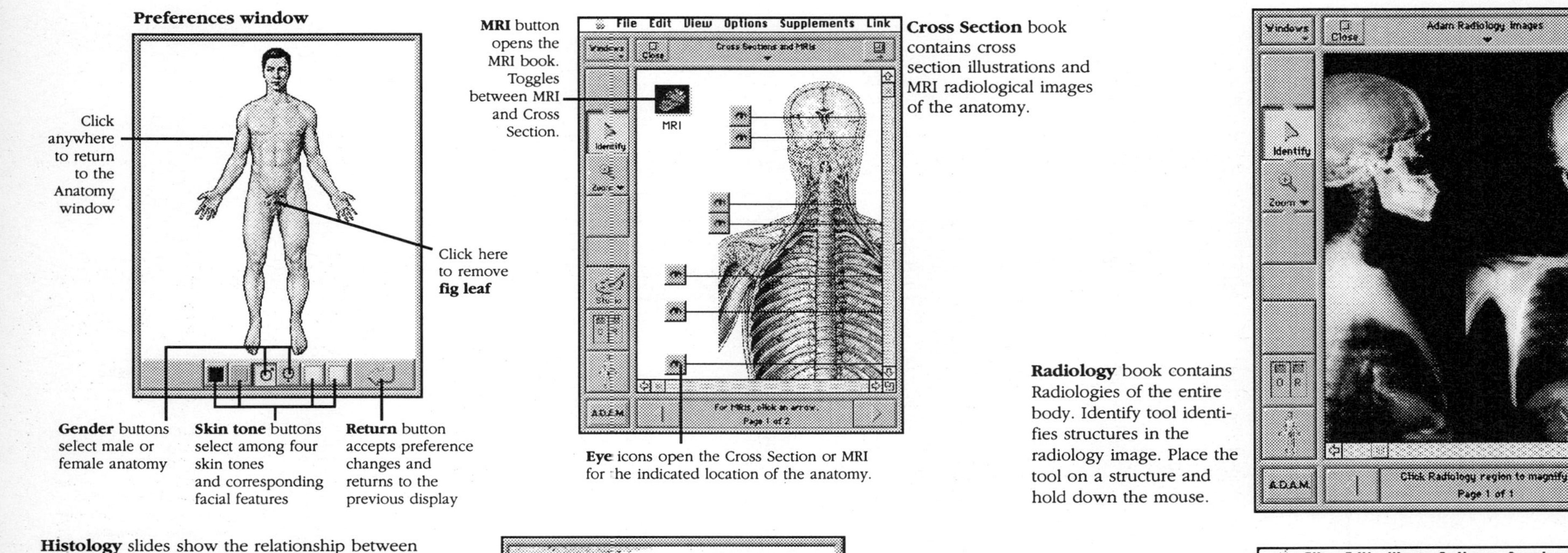

Click anywhere to return to the Anatomy window

Click here to remove **fig leaf**

Gender buttons select male or female anatomy

Skin tone buttons select among four skin tones and corresponding facial features

Return button accepts preference changes and returns to the previous display

MRI button opens the MRI book. Toggles between MRI and Cross Section.

Eye icons open the Cross Section or MRI for the indicated location of the anatomy.

Cross Section book contains cross section illustrations and MRI radiological images of the anatomy.

Radiology book contains Radiologies of the entire body. Identify tool identifies structures in the radiology image. Place the tool on a structure and hold down the mouse.

Histology slides show the relationship between histologies and gross anatomy. Histology can be accessed using the Identify tool in the anatomy window or through the Histology book on the bookshelf. Slide comparisons can be accessed through the Histology book.

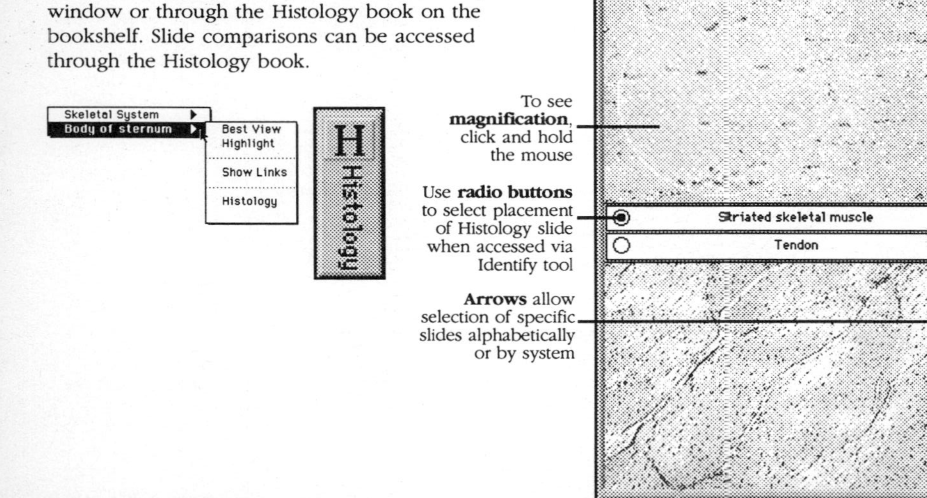

To see **magnification**, click and hold the mouse

Use **radio buttons** to select placement of Histology slide when accessed via Identify tool

Arrows allow selection of specific slides alphabetically or by system

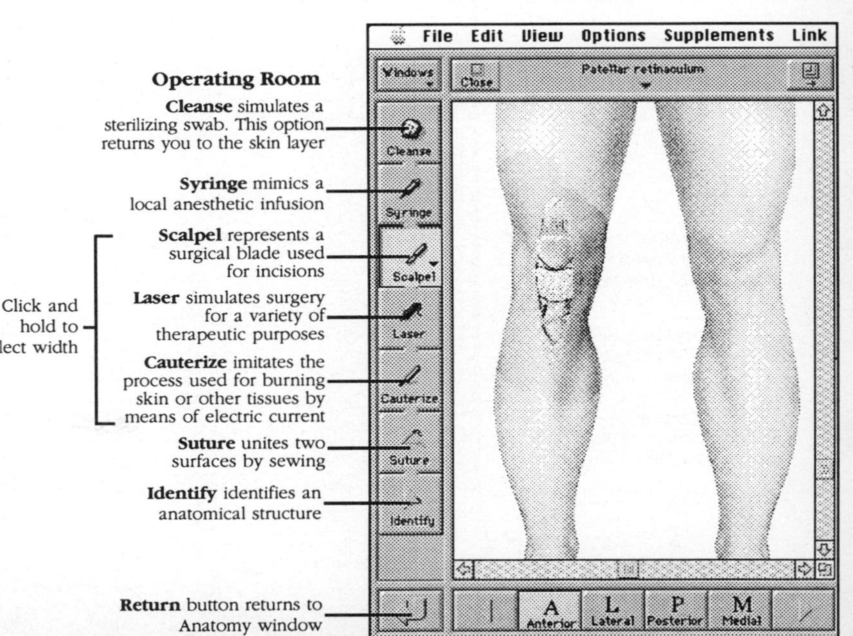

Operating Room

Cleanse simulates a sterilizing swab. This option returns you to the skin layer

Syringe mimics a local anesthetic infusion

Scalpel represents a surgical blade used for incisions

Laser simulates surgery for a variety of therapeutic purposes

Cauterize imitates the process used for burning skin or other tissues by means of electric current

Click and hold to select width

Suture unites two surfaces by sewing

Identify identifies an anatomical structure

Return button returns to Anatomy window

A.D.A.M. Standard for Windows™

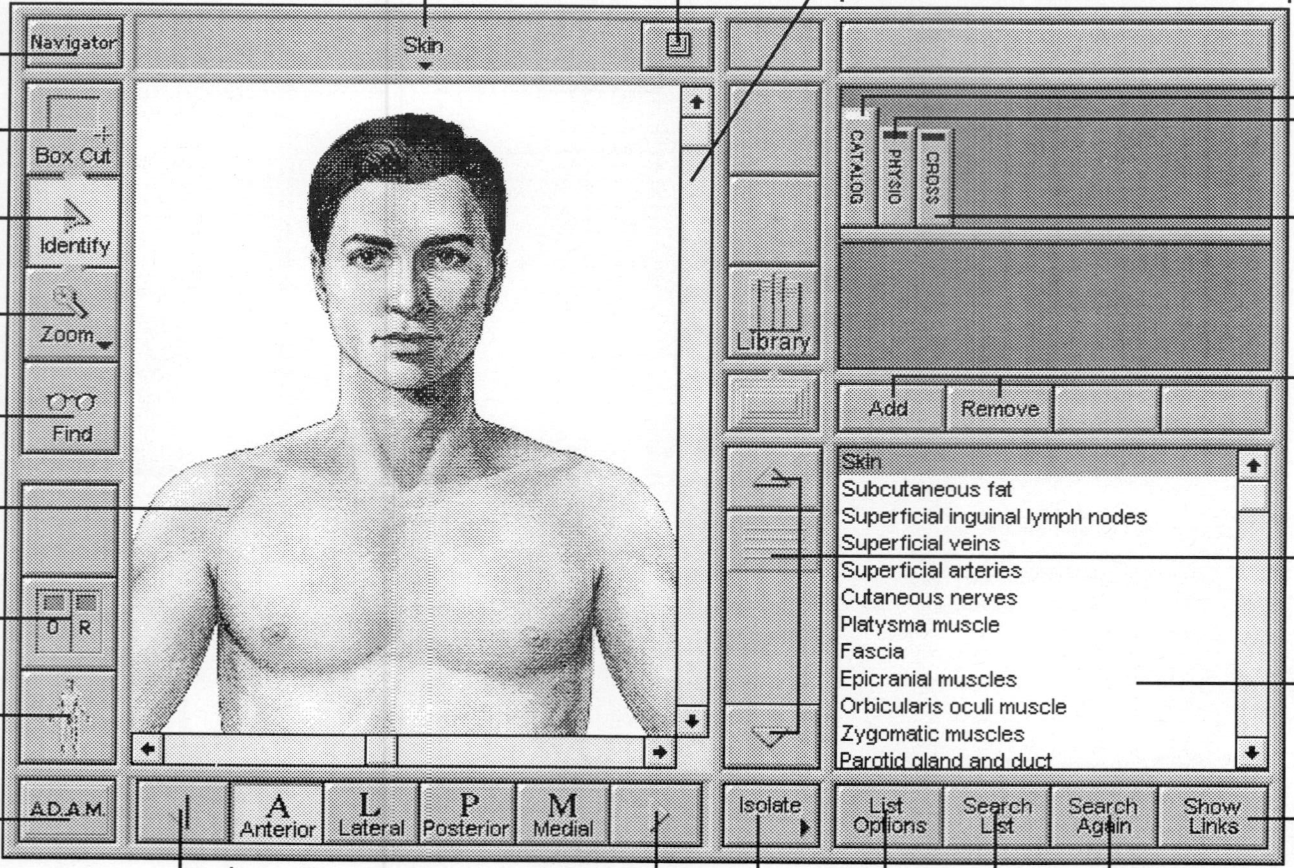
Title bar identifies contents or the selected anatomy structure. Click and hold to see the alphabetized structure list
Expand button expands the window or restores to original size
Scroll bars allow user to view anatomy not currently visible
Library window contains electronic publications
Navigator button changes the image area shown in the anatomy window
Box Cut tool defines a rectangular area to dissect
Identify tool identifies each anatomical structure and accesses related information
Zoom tool expands or condenses the anatomy image area or provides magnifying glass
Find searches for structures and substructures (if appropriate)
Anatomy window contains the anatomical image you are viewing
O.R. button opens the Operating Room (see other side)
Adam Man button scales the anatomy image to 7.4 percent of life size and restores the skin layer
A.D.A.M. accesses the A.D.A.M. Catalog of products
Catalog provides information on existing A.D.A.M. products and additional information on the Company
Physiology contains a sample of A.D.A.M Interactive Physiology modules
Cross Sections contains cross section illustrations and MRI radiological images
Add/Remove button adds and removes books on the bookshelf
Depth bar allows quick dissection through anatomical structures by dragging the bar or clicking arrows to move up or down one step at a time
Structure list window lists structures in the currently selected view
Show Links allows viewing of digitized data linked to specific structures
Left arrow moves back to the previous view
Anatomy view buttons select the anterior (front), lateral (side), posterior (back), or medial (middle) views
Right arrow moves forward to the next view
Isolate button displays options to view normal anatomy, or highlight structures
List Options allows customization of the structure list
Search List button allows searching within the structure list window
Search Again button finds the next occurrence of the name previously entered in Search List
Navigator
Skin
Box Cut
Identify
Zoom
Find
Library
CATALOG
PHYSIO
CROSS
Add
Remove
O R
A.D.A.M.
A Anterior
L Lateral
P Posterior
M Medial
Isolate
Skin
Subcutaneous fat
Superficial inguinal lymph nodes
Superficial veins
Superficial arteries
Cutaneous nerves
Platysma muscle
Fascia
Epicranial muscles
Orbicularis oculi muscle
Zygomatic muscles
Parotid gland and duct
List Options
Search List
Search Again
Show Links

Preferences

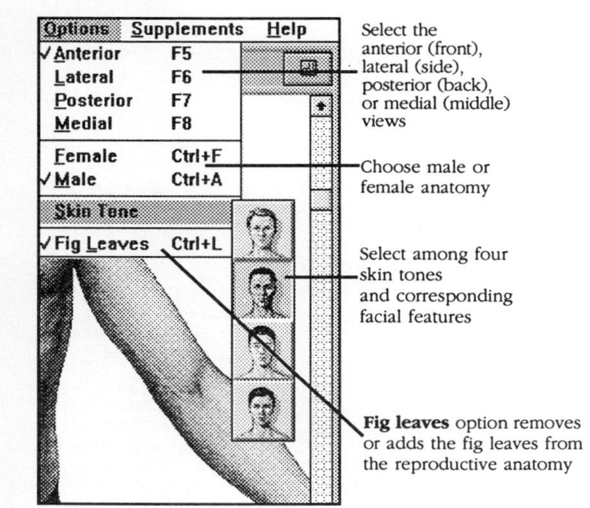

Select the anterior (front), lateral (side), posterior (back), or medial (middle) views

Choose male or female anatomy

Select among four skin tones and corresponding facial features

Fig leaves option removes or adds the fig leaves from the reproductive anatomy

Histology slides show the relationship between histologies and gross anatomy. Histology can be accessed by using the identify tool in the anatomy window.

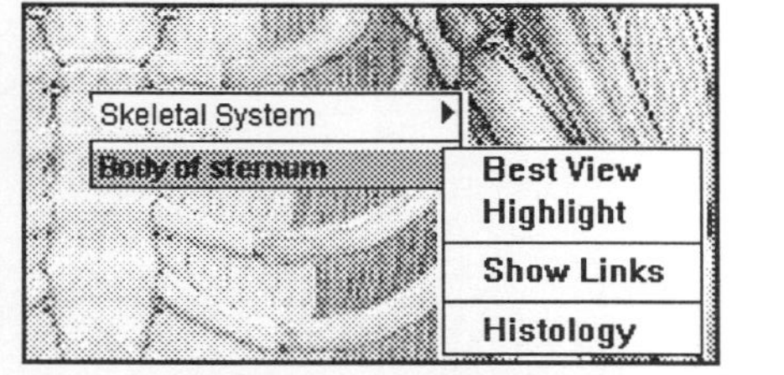

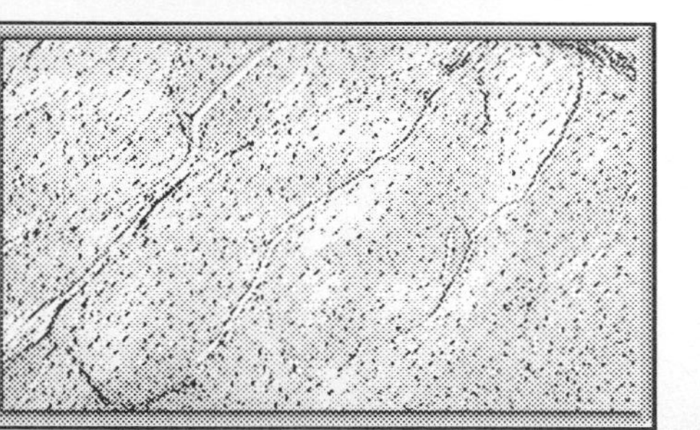

Cross Section book contains cross section illustrations and MRI radiological images that encompass the entire body. **MRIs** and Cross Sections allow for further study with the identify tool and magnifying glass.

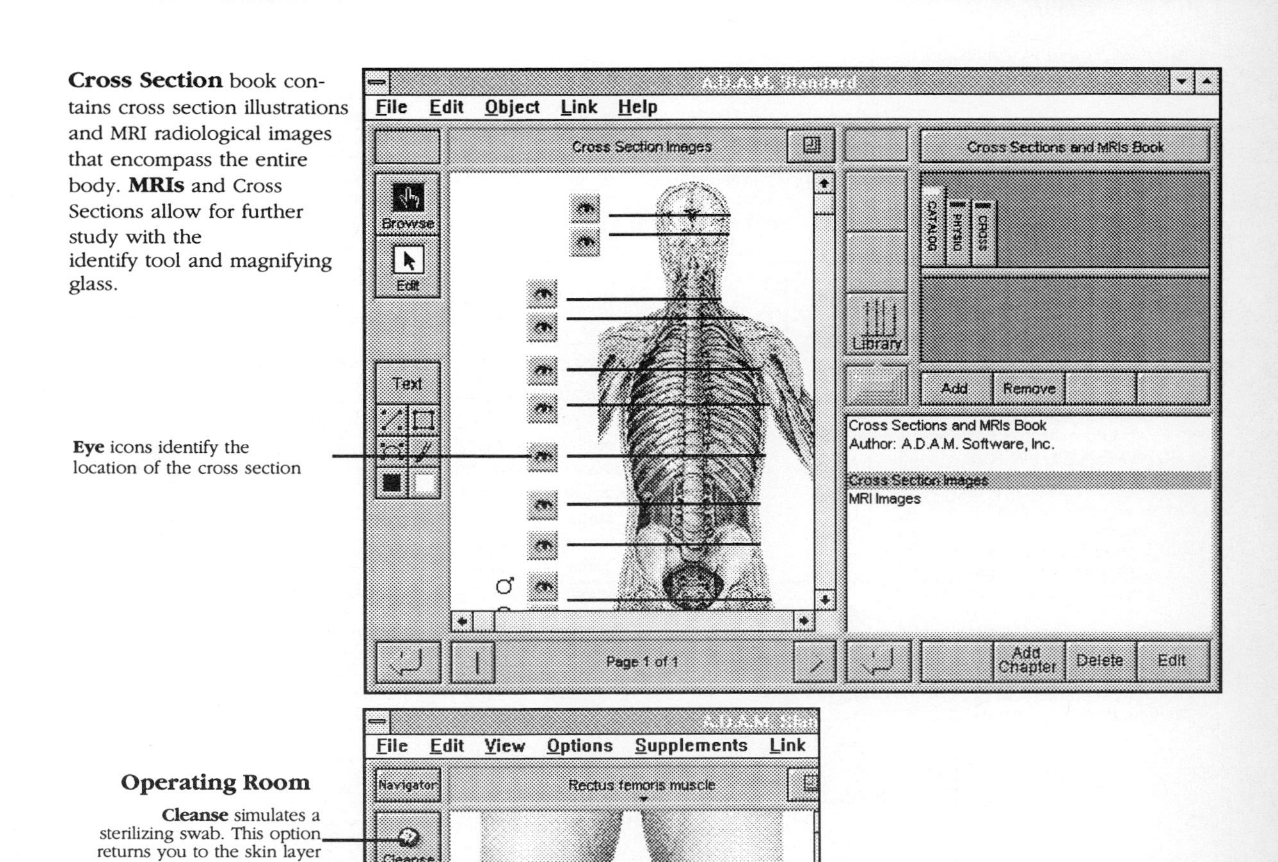

Eye icons identify the location of the cross section

Operating Room

Cleanse simulates a sterilizing swab. This option returns you to the skin layer

Scalpel represents a surgical blade used for incisions

Laser simulates surgery for a variety of therapeutic purposes

Suture unites two surfaces by sewing

Identify identifies an anatomical structure

Return button returns to anatomy window

A.D.A.M. Standard for Macintosh®

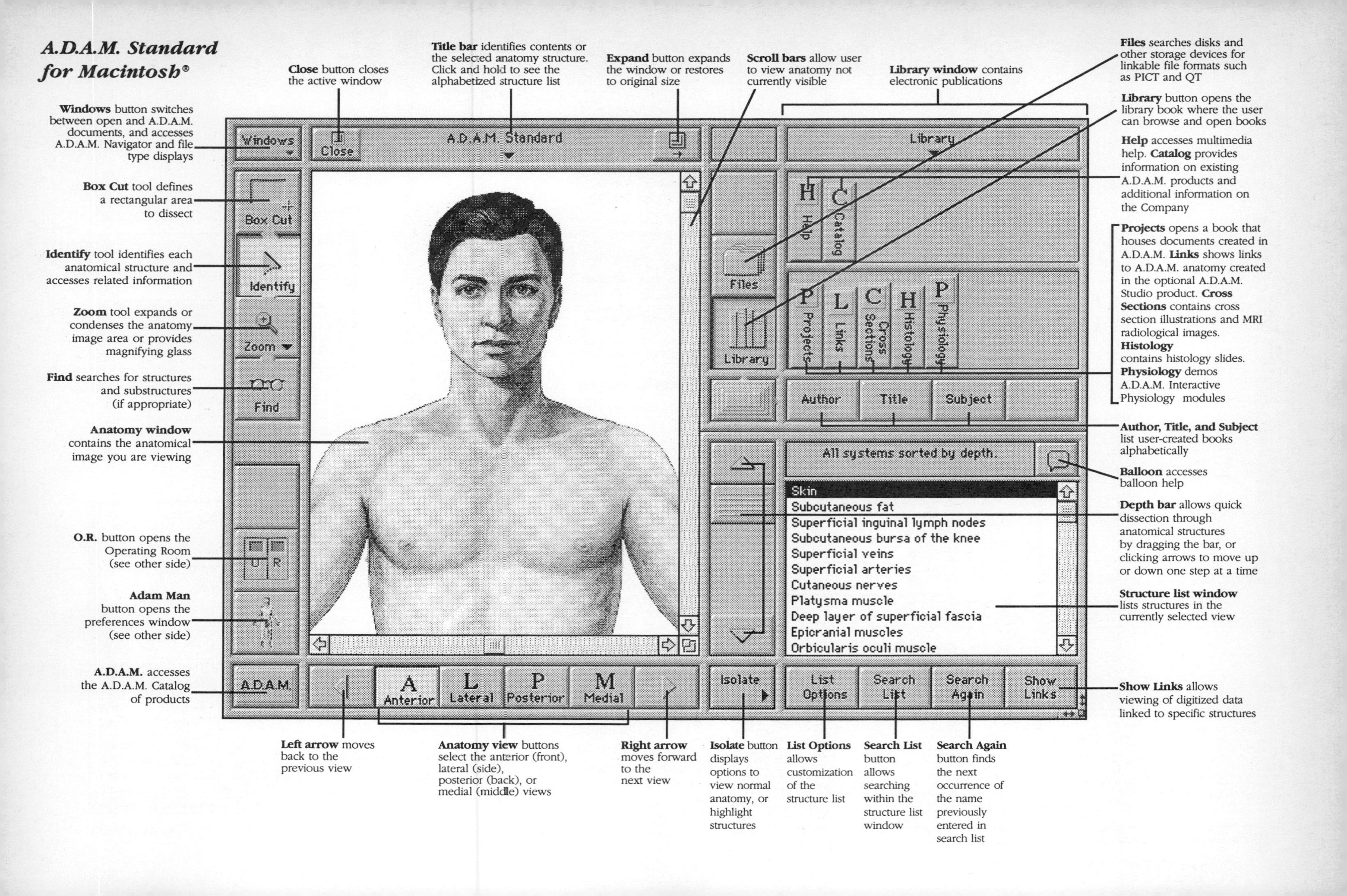

Preferences window

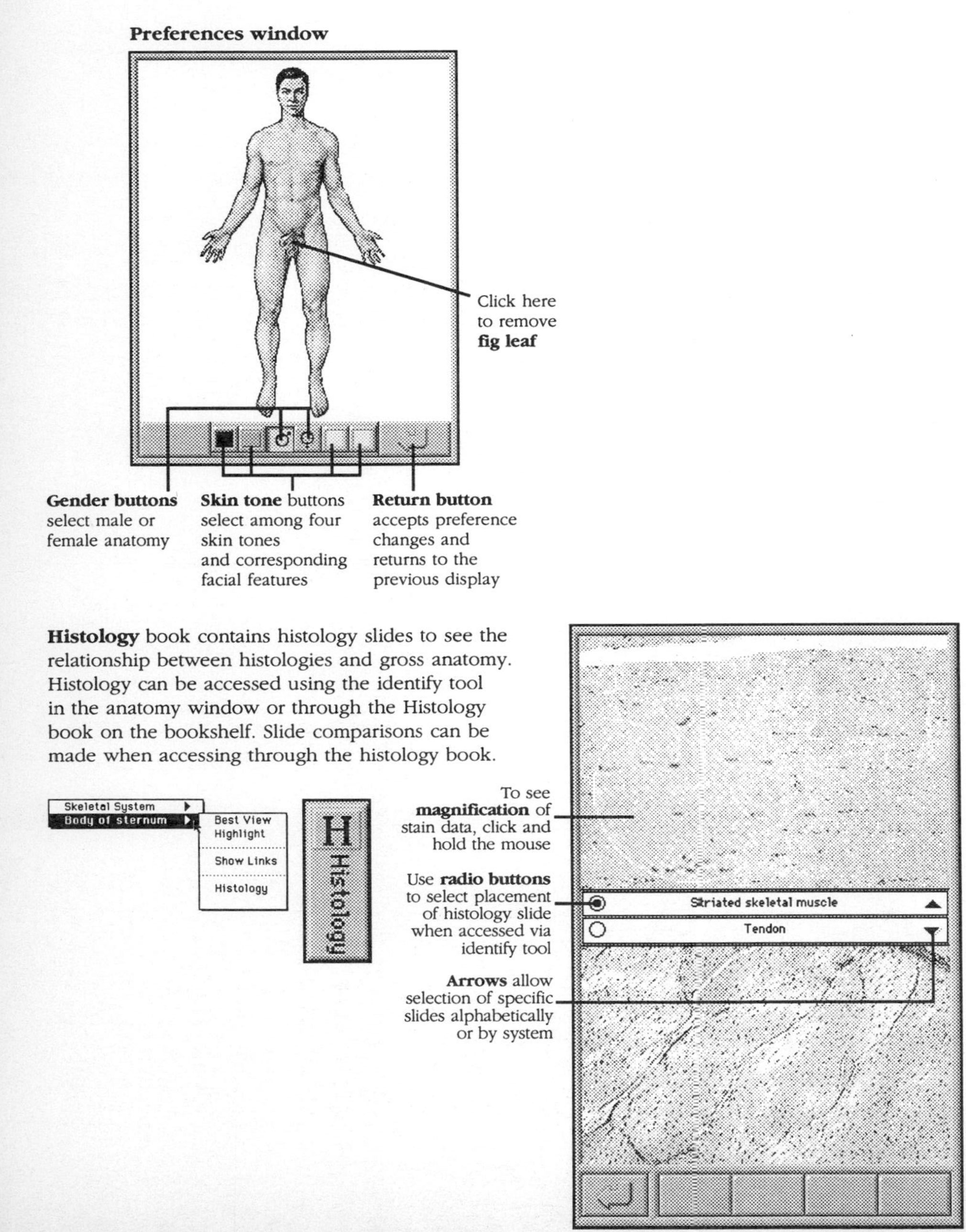

Click here to remove **fig leaf**

Gender buttons select male or female anatomy

Skin tone buttons select among four skin tones and corresponding facial features

Return button accepts preference changes and returns to the previous display

Histology book contains histology slides to see the relationship between histologies and gross anatomy. Histology can be accessed using the identify tool in the anatomy window or through the Histology book on the bookshelf. Slide comparisons can be made when accessing through the histology book.

To see **magnification** of stain data, click and hold the mouse

Use **radio buttons** to select placement of histology slide when accessed via identify tool

Arrows allow selection of specific slides alphabetically or by system

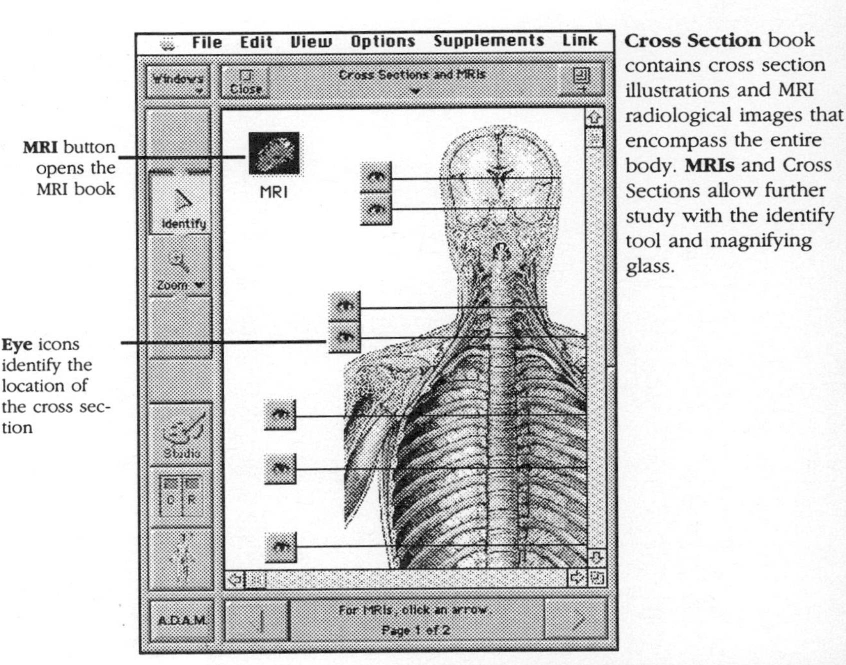

MRI button opens the MRI book

Eye icons identify the location of the cross section

Cross Section book contains cross section illustrations and MRI radiological images that encompass the entire body. **MRIs** and Cross Sections allow further study with the identify tool and magnifying glass.

Operating Room

Cleanse simulates a sterilizing swab. This option returns you to the skin layer

Syringe mimics a local anesthetic infusion

Scalpel represents a surgical blade used for incisions

Laser simulates surgery for a variety of therapeutic purposes

Cauterize imitates the process used for burning skin or other tissues by means of electric current

click and hold to select width

Suture unites two surfaces by sewing

Identify identifies an anatomical structure

Return button returns to anatomy window

1

SKELETAL SYSTEM

STUDENT OBJECTIVES

Overview

- Review basic information on the anatomy and physiology of the skeletal system.

Axial Skeleton

- Identify the bones of the skull and their markings, which can be seen in the Anterior and Lateral views.
- Describe the bony contributions to the formation of the orbit and nasal septum.
- Discuss the structure of the vertebral column and describe some of the ligaments that help stabilize the column.
- Name and identify the structures that form the bony thorax.
- Differentiate between true and false ribs.
- Describe the articulations of the ribs, both anteriorly with the sternum and posteriorly with the vertebrae.

Appendicular Skeleton

- Name the bones of the pectoral girdle and upper limb and identify their important bone markings and articulations.
- Describe the components and articulations of the pelvic girdle and identify important markings on the coxal bone.
- Name and identify the bones of the lower limb and their important markings and articulations.

Osseous Tissue

- Describe the histology of bone marrow and compact bone.

OVERVIEW

Exercise 1

Click on View in the Menu bar, drag the pointer down to the Skeletal System and release the button. Read the text overview of the skeletal system, located in the lower right side of the screen. Circle the correct answer to the following questions.

1.1 What is the total number of bones in the skeletal system?

a. 135
b. 195
c. 206
d. 236

1.2 The skeletal system does which of the following?

a. gives the body a framework
b. forms a system of levers for movement
c. acts as a site for mineral storage and blood formation
d. a, b, and c are all correct

1.3 Which of the following bones does not belong to the axial skeleton?

a. hyoid bone
b. femur
c. cervical vertebra
d. sternum

1.4 The lambdoidal suture is located between which two bones?

a. the two parietal bones
b. the parietal and frontal bones
c. the parietal and temporal bones
d. the parietal and occipital bones

1.5 The __________ vertebrae have facets that articulate with the ribs.

a. cervical vertebrae
b. lumbar vertebrae
c. thoracic vertebrae
d. sacral vertebrae

1.6 The two bones that form each pectoral girdle are the:

a. clavicle and sternum
b. clavicle and humerus
c. clavicle and scapula
d. scapula and humerus

1.7 The bone of the appendicular skeleton that provides points of attachment for the muscles of the foot and leg and increases the lateral stability of the ankle is the:

a. femur
b. metatarsal
c. tibia
d. fibula

1.8 Which of the following cell types forms new bone tissue?

a. osteoblast
b. chondroblast
c. osteocyte
d. osteoclast

1.9 The ____________ is the functional unit of bone that is sometimes referred to as the Haversian system.

a. lamella
b. canaliculus
c. osteon
d. lacuna

1.10 Yellow bone marrow fills the marrow cavity and is comprised of which of the following?

a. erythrocytes
b. fat cells
c. leukocytes
d. platelets

Exercise 2

After selecting the Skeletal System using View in the Menu bar, look at the image in the Primary window. See how many of the structures of the Skeletal System you can identify without looking at the labels. Clicking on Options from the Menu bar allows you to view the image with or without ribs. Look at the skeletal system using both options.

After quizzing yourself on your knowledge of the skeletal system, see how many structures you identified correctly by looking at the structure-identifying labels. You can choose the image to be labeled by either clicking the Label button located below the text overview or clicking on Options from the Menu bar and selecting Show Labels from the pull-down menu.

To hear pronunciations of structures of the skeletal system, first click the Identify button on the panel in the left side of the Primary window. Then select the structure you wish to hear pronounced by clicking on the structure, choosing Pronounce, and releasing the button. Practice saying the names of the structures after you have heard their correct pronunciation.

AXIAL SKELETON

The Skull

Exercise 3

Using View in the Menu bar, select Full Anatomy. The image in the Primary window should be in the Anterior view. If not, click the Anterior button below the Primary window. Click and hold the Zoom In button on the left side of the Primary window. Drag the pointer to the right and release when Zoom In becomes highlighted. You will see a magnifying glass with a (+) inside. Move the magnifying glass to the image of the head in the Primary window and click once on the area between the eyes. The head is now enlarged for better viewing of the region. Click once on the Identify button.

Now make sure that the bar above the Structure List window is labeled All Systems Sorted by Depth. If not, click the List Option button and make the necessary changes so that the structures are arranged in layers, from superficial to deep. Using the scrolling feature or the Search List button, choose the Skull layer to answer the following questions on the anterior skull.

3.1 The skull is formed from eight cranial bones and 14 facial bones. After you have identified as many of the skull bones as you can find in the Anterior view, mark a C by the bones in the list below which belong to the cranium and an F by the bones which help to form the face. You may want to use the Highlight button after you have identified each bone in order to help you to see its location more clearly.

________	nasal	________	zygomatic
________	ethmoid	________	vomer
________	maxilla	________	lacrimal
________	frontal	________	temporal
________	sphenoid	________	mandible

3.2 The orbit that surrounds the eyeball is formed from contributions of six different bones. Five of these six bones can be identified in the Anterior view. What are they?

a.

b.

c.

d.

e.

3.3 Name two separate bones which join to form the bony part of the nasal septum.

a.

b.

3.4 Identify the following bone features or markings and match them to the bone with which they are associated. You may also want to click the Lateral button located below the image in the Primary window, select the Skull layer from the Structure Layer list, and view the skull from the side before you answer all of the questions.

________	1. mastoid process	a. mandible
________	2. mental foramen	b. temporal bone
________	3. supraorbital margin	c. maxilla
________	4. infraorbital foramen	d. sphenoid bone
________	5. optic canal (foramen)	e. frontal bone
________	6. squamous portion	
________	7. ramus	
________	8. greater wing	
________	9. zygomatic process	

The Vertebral Column

Exercise 4

The vertebral column contains 26 bones and acts as a flexible support for the trunk of the body. The sacrum and coccyx in the column represent the fusion of several bones. The other 24 bones remain individual bones and are separated by intervertebral discs. Excess movements of the vertebral column are prevented by several ligaments that help to stabilize the vertebrae. Using the views and layers listed below, review the anatomy of the vertebral column. You may want to use the Zoom In button to enlarge the images.

View	Structure Layer List
Anterior	Deep Arteries
Anterior	Atlas
Posterior	Posterior vertebral column
Lateral	Skull
Medial	Skull

Answer T for true and F for false to the following statements, based on your review of the vertebral column. If the statement is false, rewrite the statement by striking out the incorrect word(s) and inserting the correct word(s).

4.1 ________ The ligamentum nuchae is located on the anterior bodies of the vertebrae in the vertebral column. *(See Anterior view, Deep Arteries layer.)*

4.2 ________ The bodies of the lumbar vertebrae are smaller than those of the thoracic vertebrae. *(See Anterior view, Atlas layer.)*

4.3 ________ Cervical vertebra C1, which articulates with the occipital condyles, is also called the axis. *(See Anterior view, Atlas layer.)*

4.4 ________ Cervical vertebrae have bifid (split) spinous processes. *(See Posterior view, Posterior Vertebral Column layer.)*

4.5 __________ The dorsal (posterior) sacral foramina lie between the median and intermediate sacral crests. *(See Posterior view, Posterior Vertebral Column layer.)*

4.6 __________ The interspinous ligament is attached to the tips of the spinous processes of the vertebrae throughout the length of the vertebral column. *(See Lateral view, Skull layer.)*

4.7 __________ Most of the thoracic vertebrae have spinous processes that project inferiorly. *(See Lateral and Medial views, Skull layer.)*

4.8 __________ The center of the intervertebral disc is the annulus fibrosus. *(See Medial view, Skull layer.)*

The Bony Thorax

Exercise 5

The bony thorax consists of the sternum, ribs, and costal cartilages. Return to the Anterior view, and click on the Windows button in the upper left corner of the Primary window. Drag the pointer down and highlight Navigator. Release the button; a little box will appear in the upper left corner of the screen with a figure that looks like the one in the current Primary window.

Drag the frame in the floating window to the Anterior Chest region. The image in the Primary window changes to the area defined in the floating window. Click the small square button in the upper left corner of the little box, and the box will disappear.

Using the Search List button below the Structure List window, search for Ribs in the list. When it is highlighted, click on Ribs and answer the following questions about the bony thorax.

5.1 Name two of the three parts of the sternum which can be seen in this view.

a.

b.

5.2 Name the other part of the sternum, which is hidden in this view.

5.3 True ribs attach directly to the sternum by their costal cartilages. False ribs either attach indirectly to the sternum by the costal cartilage of another rib or have no anterior articulation with the sternum. Describe the anterior articulation of each of the following ribs and state whether it is a true or false rib.

a. second rib:

b. fifth rib:

c. tenth rib:

Exercise 6

Each rib articulates with a thoracic vertebra of the vertebral column. In the Anterior view, select the Atlas layer from the Structure List window by using either the scrolling feature or the Search List button. The abdominal viscera have been removed and an anterior view of the vertebral column is now possible. Use the Identify button to answer the following questions.

6.1 What part of the rib articulates with the body of the thoracic vertebra?

6.2 Do some of the ribs articulate with the bodies of two adjacent vertebrae?

Now click the Posterior view button and select the Posterior Vertebral Column layer in the Structure List window.

6.3 With which part of the rib does the transverse process of the thoracic vertebra articulate?

APPENDICULAR SKELETON

The Pectoral (Shoulder) Girdle and Upper Limb

Exercise 7

The pectoral girdle is composed of the clavicle and the scapula and serves as a point of attachment for the upper limb to the axial skeleton. Many muscles that move the upper limb attach to the pectoral girdle. The bones of the upper limb include the humerus, radius, ulna, carpals, metacarpals, and phalanges. Review the anatomy of the pectoral girdle and upper limb using the Posterior and Anterior views and the layers listed below in the Structure Layer list. Type in the layer you are searching for using the Search List button and then click on it when it is highlighted.

View	Structure Layer List
Posterior	Scapula
Posterior	Humerus
Anterior	Clavicle
Anterior	Humerus

After reviewing the bones of the pectoral girdle and the upper limb in these four layers, match the bone in column B with the bone markings, bones, and other identifying features in column A. You may want to use the Find button to help you locate the structures more quickly.

	Column A	Column B
________	7.1 intertubercular groove	a. clavicle
________	7.2 proximal, middle, distal	b. scapula
________	7.3 capitate	c. humerus
________	7.4 coronoid process	d. radius
________	7.5 spine	e. ulna
________	7.6 medial epicondyle	f. carpals
________	7.7 tuberosity	g. metacarpals
________	7.8 inferior angle	h. phalanges
________	7.9 sternal end	
________	7.10 greater tubercle	
________	7.11 supraspinous fossa	
________	7.12 olecranon	
________	7.13 scaphoid	
________	7.14 coronoid fossa	
________	7.15 acromial end	

7.16 Select the Posterior view and highlight Zoom In from the Zoom button. Using the Navigator tool, select the view of the wrist and hand. Choose Humerus from the Structure Layer list. Draw and label a rough sketch of the bones of the hand in the Posterior view. The pisiform carpal bone cannot be seen in this view.

The Pelvic (Hip) Girdle

Exercise 8

The pelvic girdle is formed by two coxal (hip) bones. Each bone is sometimes referred to as an os coxae and is formed by the ossification and fusion of three separate cartilages. Using the Anterior view and the Femur layer from the Structure List window and the Posterior view and Posterior Vertebral Column layer from the Structure List window, examine the anatomy and articulations of the pelvic girdle. Use the Identify button to help you answer the following questions.

8.1 Where do the coxal bones articulate anteriorly?

8.2 With which bone of the vertebral column do the coxal bones articulate posteriorly?

8.3 What part of the coxal bone articulates with the articular cartilage of the femur?

8.4 Listed on the top row of the table below are the three separate parts of each coxal bone that have fused. Listed in the left column of the table are several bone markings that are found on each bone. Make a check mark in the box that matches the bone marking with the part of the coxal bone to which it belongs. Some of the markings may belong to more than one bone.

	Ilium	Ischium	Pubis
1. lesser sciatic notch			
2. superior ramus			
3. crest			
4. posterior inferior spine			
5. tuberosity			
6. fossa			
7. body			
8. anterior superior spine			
9. greater sciatic notch			

The Lower Limb

Exercise 9

The bones of the lower limb include the femur, patella, tibia, fibula, tarsals, metatarsals, and phalanges. Review the anatomy of the lower limb by selecting the Anterior and Posterior views and the Femur layer from the Structure List window.

9.1 Draw simple sketches of the femur, tibia, and fibula and their articulations with the hip and ankle in both Anterior and Posterior views. Label the following structures on your drawings:

Femur (Anterior view) articular cartilage covering head, neck, greater and lesser trochanters, intertrochanteric line, adductor tubercle, medial and lateral epicondyles, articular cartilage covering medial and lateral condyles

(Posterior view) articular cartilage covering head, neck, greater and lesser trochanters, intertrochanteric crest, gluteal tuberosity, linea aspera, medial supracondylar ridge, adductor tubercle, medial and lateral epicondyles, articular cartilage covering medial and lateral condyles, intercondylar fossa

Tibia (Anterior view) articular cartilage covering medial and lateral condyles, intercondylar eminence, tibial tuberosity, anterior border (crest), medial malleolus

(Posterior view) articular cartilage covering medial and lateral condyles, intercondylar eminence, soleal line, medial malleolus

Fibula (Anterior and Posterior views) head, lateral malleolus

Anterior View Posterior View

9.2 Answer the following questions about the bones of the foot after you have carefully reviewed the Anterior and Posterior views of the foot. You may want to look at the Bones of the Foot layer first in the Anterior view to see the ligaments that attach to the bones. The ligaments are removed, and the articulations of the bones in Anterior and Posterior views can be seen by choosing the Bones—Coronal Section layer in the Structure List window. You can also use the Zoom button and highlight Zoom In to enlarge the foot region.

1. What is the total number of tarsals in each foot?

2. Name them.

3. The second metatarsal articulates with which tarsal bone?

4. The cuboid bone articulates with which two metatarsal bones?
   a.

   b.

5. What is the total number of phalanges in each foot?

6. Which digit contains no middle phalanx?

OSSEOUS TISSUE

Exercise 10

Now that you know the names of the bones and many of the markings on these bones, we will review the anatomy of bone tissue. Choose Anterior view and the Bones—Coronal Section layer from the Structure Layer list. Identify the head and medullary canal of the femur. The head of the femur in an adult, as well as many other places such as the sternum and the ilium, contains red marrow and yellow marrow located between the trabeculae of spongy bone. Red bone marrow is where erythrocytes, leukocytes, and platelets are produced before they are released into the blood.

10.1 Open the Histology Book on the Library Shelf *(Macintosh users only)*. Hold down the pointer on the upward arrow in the image in the upper quarter of the screen and select By system when the pop-up menu appears. Drag the pointer down the menu and highlight Skeletal System; then choose Bone Marrow. A section of red bone marrow appears in the box in the upper quarter of the screen. The white circles represent fat tissue that has been removed during preparation, and the dark structures are the developing blood cells.

1. In a healthy individual, what percentage of the red bone marrow would you estimate to be fat and what percentage do you think is made up of developing cells?

2. Draw a rough sketch of a histological section of red bone marrow and label the fat and blood cells.

10.2 The medullary cavity is the inside of the shaft of a long bone and contains primarily fat and other connective tissue cells. It is surrounded by compact bone tissue. Repeat the procedure for selecting the section as you did in Exercise 10.1, except choose Ground bone from the Skeletal System. Osteocytes (bone cells), which are the darkly stained structures in this preparation, are arranged concentrically within the bone matrix to form the structural unit of compact bone called the osteon or Haversian system. Draw a simple sketch of several of these units, as seen in this histological section. After you have completed this exercise, close the Histology Book by clicking the Return arrow located below the lower section.

2

JOINTS

STUDENT OBJECTIVES

Overview

- Review basic information on the anatomy and physiology of joints.
- Classify joints both structurally and functionally.

Fibrous Joints

- Describe the structure of fibrous joints.
- Name and identify three examples of fibrous joints.

Cartilaginous Joints

- Differentiate between a synchrondrosis and a symphysis joint.
- Name and identify three examples of cartilaginous joints.

Synovial Joints

- Describe the anatomy of the shoulder, elbow, hip, and knee joints. Identify the capsule surrounding each joint and the major ligaments and muscle tendons which help to reinforce and stabilize the joints.
- Differentiate between a bursa and a synovial tendon sheath.

OVERVIEW

Exercise 1

Click on View in the Menu bar and select the Skeletal System. Scroll through the text overview in the lower right side of the screen until you reach the text on joints (Section 6). Read the overview and answer the following questions by circling the correct answer.

1.1 An example of a freely movable joint is a ________________.

a. synarthrosis
b. symphysis
c. synovial joint (diarthrosis)
d. synchondrosis

1.2 Which of the following are parts of a synovial joint?

a. joint cavity
b. articular cartilage
c. synovial membrane and fluid
d. a, b, and c are all correct

1.3 What covers the ends of bones to prevent excess wear and tear during movement?

a. fibrous tissue
b. articular cartilage
c. fat
d. muscular tissue

1.4 The type of synovial joint in which the articulating bones allow movement along one plane, similar to a door, is classified as a ____________ joint.

a. condyloid
b. pivot
c. ball-and-socket
d. hinge

1.5 In a ________ joint, one round-shaped articulating bone fits within a corresponding depression on another bone.

a. ball-and-socket
b. condyloid
c. pivot
d. plane

1.6 An example of a condyloid joint is the:
a. hip joint
b. metacarpophalangeal joint
c. elbow joint
d. radioulnar joint

1.7 The types of joints that allow short, gliding movements of the bones are:
a. saddle joints
b. plane joints
c. ball-and-socket joints
d. condyloid joints

1.8 The carpometacarpal joint of the thumb is an example of a ________________.
a. condyloid joint
b. hinge joint
c. pivot joint
d. saddle joint

1.9 An example of a symphysis joint is the:
a. radioulnar joint
b. knee
c. carpometacarpal joint
d. intervertebral disc

1.10 What connective tissue structure attaches muscle to bone?
a. ligament
b. disc
c. tendon
d. cartilage

Exercise 2

Joints are classified structurally and functionally. Their structural classification is based on the presence or absence of a cavity between the bones and the type of tissue that fills in the cavity if there is no space. These joints are classified as fibrous, cartilaginous, or synovial joints. Functional classification of joints is determined by the degree of movement that is permitted between the bones and is classified as a synarthrosis, amphiarthrosis, or diarthrosis. After reading the overview on joints, you should be able to classify the joints in the following table both structurally and functionally. The information that you use to complete the table will help you to answer other questions in this chapter.

Type of Joint	Structure	Function
2.1 suture		
2.2 gomphosis		
2.3 syndesmosis		
2.4 symphysis		
2.5 ball-and-socket		
2.6 hinge		

FIBROUS JOINTS

Fibrous joints are joints in which bones are connected by fibrous connective tissue of varying lengths. They are functionally classified as either immovable or slightly movable, depending on the length of fiber that connects the two bones. You may want to refer to the text overview to help you answer some of the questions in the following exercises.

Exercise 3

Click on View in the Menu bar and select Full Anatomy. Using the Lateral view, select the Skull layer from the Structure List window to review the structure of one type of joint in which the bones are connected by short, fibrous connective tissue.

3.1 What type of fibrous joints occur between the bones of the skull?

3.2 Identify the frontal, parietal, occipital, and temporal bones. Refer to the text overview or your textbook to help you name the fibrous joints that are located between the bones. They are not labeled in this view.

a. joint located between the frontal and parietal bones:

b. joint located between the parietal and temporal bones:

c. joint located between the parietal and occipital bones:

d. joint located between the two parietal bones:

3.3 Name another example of an immovable joint that you can see in this lateral view of the skull. *(Hint: Look at the maxilla and mandible.)*

3.4 What functional classification term describes immovable joints?

Exercise 4

Another type of fibrous joint connects bones with longer fibrous tissues or sheets of tissues and may allow slight movement between the bones. These types of joints are called syndesmoses. Choose Anterior view and select the Radius layer from the Structure List window to see an example of the syndesmosis that occurs between the radius and ulna. This type of structure also is located between the tibia and fibula.

4.1 What is the name of this structure?

4.2 What is the functional classification term used to describe slight movement which is allowed between these bones?

CARTILAGINOUS JOINTS

Cartilaginous joints are those in which bones are connected by either hyaline cartilage or fibrocartilage.

Exercise 5

A synchondrosis joint is one in which the bones are united by hyaline cartilage. Using Anterior view, select the Ribs layer in the Structure List window. Identify the manubrium of the sternum, costal cartilage, and the first rib. The hyaline costal cartilage connecting the two bones is an example of a synchondrosis type of joint.

5.1 What functional classification term describes the type of movement possible at this joint?

Exercise 6

Another type of cartilaginous joint is a symphysis joint. In this type of joint, bones are united by a plate of fibrocartilage.

6.1 Look at the vertebral column by selecting the Medial view and choosing the Skin layer from the Structure List window. An example of a symphysis joint can be identified between the bodies of adjacent vertebrae in the vertebral column.

a. What are these structures which lie between adjacent vertebrae and act as shock absorbers to help protect the vertebrae?

b. Each one of them is composed of two parts. Name their components.

6.2 Name another example of a symphysis joint that can be seen when selecting the Anterior view and the Femur layer from the Structure List window. (*Hint: Look in the region of the anterior pelvic girdle.*)

6.3 What type of movement is allowed at these two examples of symphysis joints examined in Exercises 6.1 and 6.2?

SYNOVIAL JOINTS

A synovial joint occurs between bones separated by a fluid-filled cavity. This type of joint is classified as a freely movable joint, or a diarthrosis. The synovial membrane-lined fibrous capsule that surrounds the joint is usually stabilized and strengthened by ligaments outside (extracapsular) or inside the capsule (intracapsular). Examples of synovial joints include the shoulder, elbow, hip, and knee joints.

Exercise 7

Using the Anterior view, select the Fibrous Joint Capsule of the Shoulder and Elbow layer in the Structure List window. You may want to use the Zoom In button to enlarge the region before answering the following questions.

7.1 Identify the fibrous capsule surrounding the shoulder joint. Ligaments help to reinforce the capsule surrounding the joint. Two of these ligaments can be seen near the superior part of the capsule. What are they?

a.

b.

7.2 In addition to ligaments, the muscle tendons of the "rotator cuff" muscles (supraspinatus, infraspinatus, teres minor, and subscapularis) also help to stabilize the capsule of the joint. Which two of these muscles can you see in this Anterior view?

a.

b.

7.3 Bursae are friction-reducing "fluid-filled sacs" that lie between bones and muscles, tendons, ligaments, and skin. Identify the bursa near the superior part of the capsule of the shoulder joint. Name the two muscle tendons that it separates.

a.

b.

7.4 Another structure, which is like a bursa but is usually elongated and wraps around a tendon, is a synovial tendon sheath. Name the muscle whose tendon sheath appears to go inside the fibrous capsule of the shoulder joint.

Exercise 8

The anatomy of the elbow joint can also be reviewed using the same view and layer as in Exercise 7.

8.1 Identify the fibrous capsule surrounding the joint, and name the two collateral ligaments that lie outside the capsule.

a.

b.

8.2 What does the annular ligament surround?

Exercise 9

The anatomy of the hip joint can be reviewed using the Anterior view, Obturator Nerve layer, and the Posterior view, Ligaments of the Hip layer in the Structure List window.

9.1 Name three extracapsular ligaments that help to reinforce the fibrous capsule of the hip joint, and describe their attachments to the femur and coxal bone. Two of these ligaments can be seen in the Anterior view, and the other can be seen in the Posterior view.

a.

b.

c.

9.2 When the fibrous capsule is removed by "dissecting" (scrolling) down and selecting the Synovial Joint Capsule of the Hip layer in both the Anterior and Posterior views, the location of the synovial lining of the fibrous capsule can be seen covering the joint. Now choose the Anterior view and then the Posterior view and select the Femur layers in both views in the Structure List window to answer the following questions.

a. What covers the head of the femur where it articulates with the coxal bone?

b. What is the name of the circular rim of fibrocartilage that deepens the acetabulum where the head of the femur articulates with the coxal bone?

Exercise 10

The knee joint is the largest joint in the body. As in other joints, muscle tendons and ligaments help to stabilize the joint, and it is protected by fat pads and bursae. Using the Anterior view and selecting the Rectus Femoris Muscle layer in the Structure List window, review the external anatomy of the knee joint.

10.1 The capsule of the knee only partially covers the joint. Where the capsule is absent on the anterior of the joint, the patellar ligament and the medial and lateral patellar retinacula are located. Identify these three structures. These structures are continuations of the tendon of the quadriceps femoris muscle. Name the three muscles that can be seen in this anterior view whose tendons help to form the quadriceps femoris tendon.

a.

b.

c.

10.2 Name the other muscle of the quadriceps femoris group that cannot be seen in this view.

10.3 What "yellow" structure lies inferior to the patella and posterior to the patellar ligament?

10.4 Name three muscle tendons that help to stabilize the medial side of the knee and insert on the tibia near the tibial tuberosity.

a.

b.

c.

Exercise 11

Select the Patella layer from the Structure List window. With many of the muscle tendons removed, you can now review some of the extracapsular and intracapsular ligaments of the knee.

11.1 Name two extracapsular ligaments that can be identified in this layer and describe where they attach on each bone.

a.

b.

11.2 Name the intracapsular ligament that can be seen in this view.

11.3 What blood vessels supply and drain the knee joint?

Exercise 12

Change to the Posterior view and select the Posterior Cruciate Ligament layer from the Structure List window.

12.1 Draw and label a posterior view of the knee joint. Include in your drawing the following structures: articular cartilage of the femoral and tibial condyles, tibial and fibular collateral ligaments, anterior and posterior cruciate ligaments, and medial and lateral menisci.

12.2 Describe the relationship of the medial meniscus to the tibial collateral ligament.

3

MUSCULAR SYSTEM

STUDENT OBJECTIVES

Overview

- Review basic information on the anatomy and physiology of the muscular system.
- Compare the histology of skeletal and cardiac muscle tissue based on the arrangement of the nuclei and fibers.

Major Skeletal Muscles of the Body

- Identify and describe the location of the major skeletal muscles of the head, anterior and anterolateral neck, posterior neck and back, anterior thorax, anterior and lateral abdominal wall, anterior and posterior arm, anterior and posterior forearm, anterior and medial thigh, posterior hip and thigh, and anterior, lateral, and posterior leg.
- Identify the muscles in a cross-sectional drawing of the mid forearm and mid leg.

OVERVIEW

Exercise 1

Click on View in the Menu bar and select the Muscular System. Read the text overview and answer the following true/false questions about the Muscular System. Mark T if the statement is true and F if the statement is false. Rewrite the false statements by striking out the incorrect word(s) and inserting the correct word(s).

1.1 _____ The body contains 206 skeletal muscles, which account for about 10% of body weight.

1.2 _____ Cardiac muscle is involuntary and is composed of striated muscle fibers.

1.3 _____ Smooth muscle in the body is located in the heart, visceral organs, and the walls of blood vessels.

1.4 _____ The type of muscle responsible for peristaltic contractions in the digestive tract is skeletal muscle.

1.5 _____ The connective tissue covering that surrounds a fasicle of muscle fibers is the epimysium.

1.6 _____ A muscle fiber is composed of many long cylindrical structures called myofibrils.

1.7 _____ Thick myofilaments contained within myofibrils are composed of actin protein.

1.8 _____ The functional unit of skeletal muscle is the sarcomere.

1.9 _____ Strong sheets of connective tissue that connect muscles to bones are ligaments.

1.10 _____ Good muscle tone is important in helping to maintain body posture.

1.11 _____ Weightlifting, sprinting, and push-ups are types of exercises that use slow-twitch muscle fibers.

Exercise 2

The histological differences between striated skeletal and cardiac muscle tissue can be seen by opening the Histology Book on the Library Shelf (Macintosh users only). Click on the Up arrow in the lower right corner of the upper box. Drag the pointer down to the Muscular System and select Striated Skeletal Muscle from the choices. A histological section of striated skeletal muscle will appear in the upper box. Now click the Down arrow in the upper right corner of the lower box and select Striated cardiac muscle. After comparing the two types of striated muscle, describe the differences based on the arrangement of the nuclei and fibers seen in these sections.

After you have completed this exercise, click the Return arrow below the lower box and return to the overview of the muscular system.

Exercise 3

You may already know many of the muscles that are shown in the Primary window. Quiz yourself without looking at the labels. If you need help, click the Label button and the names will appear. You can also listen to the correct pronunciation of the name if you need to.

3.1 How many muscles of the body did you identify before looking at the labels?

MUSCLES OF THE HEAD

Exercise 4

The muscles of facial expression can be seen when you choose Full Anatomy using View in the Menu bar. In Anterior view, examine the Platysma layer and the Epicranial Muscles layer in the Structure List window. Match the muscles of the face and scalp with their descriptions. You may want to use the Zoom button to magnify the region.

		Description	Muscle
4.1	_____	encircles the mouth	a. zygomaticus major
4.2	_____	located between the eyes	b. platysma
4.3	_____	originates on the zygomatic bone and inserts on the corner of the mouth	c. nasalis
			d. frontalis
4.4	_____	elevates the upper lip	e. mentalis
4.5	_____	superficial muscle of the neck	f. procerus
4.6	_____	surrounds the orbit	g. levator labii superioris
4.7	_____	located on the bridge of the nose	h. orbicularis oculi
4.8	_____	depresses the lower lip	i. depressor labii inferioris
4.9	_____	muscle of the anterior forehead	j. orbicularis oris
4.10	_____	located near the midline of the mandible	

Exercise 5

The muscles described in Exercise 4 and others can also be identified in the Lateral view and the Zygomatic Muscles layer in the Structure List window.

5.1 Name the two muscles that constitute the epicranius muscle.

a.

b.

5.2 Name the cranial aponeurosis that connects the two muscles described in 5.1.

5.3 What is the deep muscle of the cheek which is penetrated by the parotid duct?

5.4 Which muscle of mastication overlies the angle of the mandible and is crossed by the parotid duct?

MUSCLES OF THE ANTERIOR AND ANTEROLATERAL NECK

Exercise 6

Underlying the platysma muscle are several layers of muscles located on the anterior and anterolateral neck. Some of these muscles have insertions on the hyoid bone and are classified as infrahyoid or suprahyoid muscles, depending on whether they are inferior or superior to the hyoid. First choose View in the Menu bar and select Muscular System. After completing your review of the muscles of the neck, return to Full Anatomy using View in the Menu bar. Select Anterior view and examine the Sternocleidomastoid Muscle layer, Sternohyoid Muscle layer, and the Sternothyroid Muscle layer from the Structure List window. Answer the following questions about the origins and insertions of the muscles of the neck based on your observations.

6.1 Which of the infrahyoid muscles has its origin from the scapula?

6.2 Name the muscle that has two heads of origin from the sternum and inserts on the mastoid bone.

6.3 What muscle is located medial to the omohyoid muscle and inserts on the hyoid bone?

6.4 Name the muscles that originate from cervical vertebrae and insert on the first rib.

6.5 Which belly of the digastric can be seen inserting on the hyoid bone in these views?

6.6 Name the infrahyoid muscle that originates from the sternum and underlies the sternohyoid muscle.

6.7 Which suprahyoid muscle crosses over the posterior belly of the digastric to insert on the hyoid bone?

MUSCLES OF THE POSTERIOR NECK AND BACK

Exercise 7

The posterior neck and back have many muscles which have several different actions. Some of the superficial muscles have actions on the head, pectoral girdle, and humerus. Other deeper back muscles are important for breathing, and even deeper layers of back muscles function to maintain body posture. In the following exercises, you are going to travel from superficial to deep and review the muscles as they are layered in the neck and back.

Begin your review of the superficial muscles of the neck and back by selecting the Posterior view and the Trapezius Muscle layer from the Structure List window. After you have completed your review of this superficial layer, "dissect" a little deeper by choosing the Rhomboideus Muscle layer. Name the muscles that are described below.

7.1 ______________ Superficial muscle of the back that originates from the head and cervical and thoracic vertebrae.

7.2 ______________ Muscle that lies in the supraspinous fossa.

7.3 ______________ Muscle originating on the inferior angle of the scapula and inserting on the humerus.

7.4 ______________ Superficial muscle of the back that "wraps" around the lower trunk and inserts on the humerus.

7.5 ______________ Muscle of the posterior neck that travels laterally from the lower cervical spines to the mastoid process.

7.6 ______________ Muscle that arises on thoracic spines and inserts on most of the medial border of the scapula.

7.7 ______________ Muscle that inserts on the superior angle of the scapula and originates from cervical transverse processes.

7.8 ______________ Muscle located deep in the splenius capitus, is medial in position.

7.9 ______________ Superficial muscle that "caps" the shoulder.

7.10 ______________ Muscle located between the infraspinatus and teres major that inserts on the humerus.

7.11 ______________ Muscle that inserts on the scapula near the levator scapulae and originates from the spines of C7 and T1.

7.12 ______________ Muscle that originates below the spine of the scapula and inserts on the humerus.

Exercise 8

Continue your examination of the posterior muscles by choosing the Serratus Posterior muscle layer. After you have reviewed the muscles that can be seen clearly in this view, select the Iliocostalis Lumborum Muscle layer and examine the deep muscles of the back. Answer the following questions based on your review of these two layers.

8.1 Which bone of the pectoral girdle has been removed so that you can see the deeper muscles of the posterior back?

8.2 Name the muscle that lies on the anterior surface of the bone described in 8.1, which is now visible.

8.3 Name the two posterior respiratory muscles that arise on the spines of the thoracic and lumbar vertebrae and insert on the ribs.

a.

b.

8.4 Three columns of deep back muscles constitute the erector spinae, or sacrospinalis muscles. List the three groups of muscles and describe their position relative to the midline of the back.

a.

b.

c.

8.5 What is the function of this large muscle mass?

Exercise 9

The deepest back muscles have attachments between adjacent vertebrae and between vertebrae and ribs, and can be seen when the erector spinae muscles are removed. Review the muscles in the following layers: Iliocostalis cervicis muscle, Interspinalis muscle, Semispinalis muscle, Multifidus muscle, and Rotatores muscle. You can see how many layers of muscles are located in the back when you study each layer carefully.

9.1 List the attachments of each of the following deep back muscles.

a. semispinalis cervicis and thoracis

b. multifidus

c. rotatores

d. levator costae

e. interspinalis

f. intertransversarius

9.2 You can also see the deep muscles of the posterior neck in the Rotatores Muscle layer. What are they?

a.

b.

c.

d.

MUSCLES OF THE ANTERIOR THORAX

Exercise 10

Superficial muscles of the anterior thorax attach to bones of the pectoral girdle and the humerus and produce movements of the scapula and clavicle and of the arm at the shoulder joint. Deeper muscles of the anterior thorax are involved with breathing and lie between adjacent ribs. Begin your study by choosing the Anterior view and selecting the Pectoralis Major Muscle layer from the Structure List window. After you have reviewed the origin and insertion of the pectoralis major, examine the origin and insertion of the subclavius and pectoralis minor in the Pectoralis Minor Muscle layer. Finally, choose the Serratus Anterior Muscle layer to review.

Because muscle actions are designed to draw the origin of a bone closer to its insertion, you should be able to determine the actions of a muscle once you locate its attachments. List the actions of each of the following muscles based on your observations of their origins, insertions, and locations on the anterior thorax.

10.1 pectoralis major

10.2 subclavius

10.3 pectoralis minor

10.4 serratus anterior

Exercise 11

Deep muscles of the thorax are important in movements of the thorax during breathing. Select the External Intercostal Muscle layer from the Structure List window. Locate the first layer of deep muscles—the external intercostal muscles.

11.1 What is the orientation of the external intercostal muscle fibers between adjacent ribs?

11.2 Where are the muscle fibers located on the thorax?

11.3 Name the "white" structure that continues the muscle fibers to the sternum.

Exercise 12

Now you are going to perform surgery and look at the second and third layers of deep thoracic muscles. Click the OR button and open the doors to the operating room. Select the Scalpel button and select Wide for the incision width. Make an incision from the middle of the first rib to the middle of the tenth rib. Repeat this procedure, layer by layer, and review all the structures in each layer until you have reached the Innermost Intercostal Muscle layer.

12.1 Name the second layer of muscles of the deep anterior thorax that underlie the external intercostal muscles.

12.2 Describe the orientation of the fibers of the muscle described in 12.1.

12.3 Do the muscle fibers described in the questions above reach the sternum?

12.4 Name the two muscles that form the third layer of muscles.

a.

b.

12.5 Name the structure that connects the two muscles in 12.4.

12.6 Between which two layers of muscles do the intercostal arteries, veins, and nerves travel?

12.7 Which vessels lie lateral to the transversus thoracis muscles?

MUSCLES OF THE ANTERIOR AND LATERAL ABDOMINAL WALL

Exercise 13

The muscles of the anterior and lateral abdominal wall flex and laterally bend the vertebral column and protect the internal viscera. Complete the paragraph below after you have reviewed the following layers in Anterior view: External Abdominal Oblique Muscle, Internal Abdominal Oblique Muscle, and Rectus Abdominis Muscle.

The most superficial anterolateral abdominal muscle is the (13.1) ________________. Its muscle fibers insert into its (13.2) ________________, which joins at the midline with the layer from the other side of the abdomen, at the (13.3) ________________. The (13.4) ____________ muscle lies underneath the external oblique and has fibers that run at right angles to the overlying muscle. The deepest muscle of the anterolateral abdominal wall is the (13.5) ________________. Its fibers are oriented (13.6) ________________ across the abdomen. The rectus abdominis originates from the (13.7) ________________ and inserts on the (13.8) ________________. Several (13.9) ________________ reinforce the segments of the rectus abdominis on its anterior surface. The muscle that inserts on the linea alba and draws it inferiorly toward the pubis is the (13.10) ________________. Intercostal nerves and blood vessels travel between the (13.11) ________________ and (13.12) ________________ layers of muscle on the anterolateral abdominal wall to supply structures in the region.

MUSCLES OF THE ANTERIOR AND POSTERIOR ARM

Exercise 14

Muscles located on the anterior and posterior arm cross the elbow joint and produce flexion and extension of the forearm, respectively. The muscles of the anterior arm can be seen by choosing the Anterior view and the Biceps Brachii Muscle layer from the Structure List window. The muscles of the deeper anterior layer are shown in the Brachioradialis Muscle layer.

14.1 Which head of the biceps brachii originates on the coracoid process of the scapula?

14.2 Name another muscle that also originates on the coracoid process but inserts on the humerus and does not cross the elbow joint.

14.3 Where is the bicipital aponeurosis?

14.4 Name the forearm flexor that lies deep in the biceps brachii.

14.5 What is the insertion of the brachioradialis?

14.6 Does the brachioradialis flex or extend the forearm at the elbow joint?

Exercise 15

The extensors of the forearm can be seen when you select the Posterior view and the Triceps Muscle layer.

15.1 Which head of the triceps brachii muscle originates on the scapula and crosses the teres major muscle?

15.2 Name another muscle that crosses the elbow joint and assists the triceps brachii in extension of the forearm.

MUSCLES OF THE ANTERIOR AND POSTERIOR FOREARM

Exercise 16

Muscles that overlie the anterior and posterior forearm produce flexion and extension of the wrist and fingers. Choose the Anterior view, and look at the following layers to see the muscles that cause these movements: Brachioradialis Muscle, Flexor Carpi Radialis Muscle, Flexor Digitorum Superficialis Muscle, Flexor Digitorum Profundus Muscle, and Pronator Quadratus Muscle. After reviewing the muscles of these superficial and deep layers, match the muscle of the anterior forearm to its description.

		Description	Muscle
16.1	_____	originates medially and crosses forearm superior to insert on lateral radius	a. flexor carpi radialis
			b. flexor pollicis longus
16.2	_____	inserts by its palmar aponeurosis	c. pronator teres
16.3	_____	flexes wrist and adducts hand	d. flexor digitorum profundus
16.4	_____	superficial flexor of the fingers	e. palmaris longus
16.5	_____	deep flexor of the thumb	f. flexor capri ulnaris
16.6	_____	originates on ulna and inserts on the radius	g. pronator quadratus
			h. flexor digitorum superficialis
16.7	_____	flexes wrist and abducts hand	
16.8	_____	lumbricals originate from the tendons of this muscle	

Exercise 17

Extension movements of the wrist and fingers are caused by muscles of the posterior forearm. These muscles are located in superficial and deep compartments similar to the flexor compartments on the anterior forearm. Choose the Posterior view and review the following layers from the Structure List window: Triceps Muscle, Extensor Carpi Ulnaris Muscle, Extensor Digiti Minimi Muscle, Extensor Carpi Radialis Brevis Muscle and Supinator Muscle.

17.1 Name the structure at the wrist that prevents extensor tendons from "bowstringing."

17.2 List three extensor muscles of the wrist and fingers that are located in the superficial compartment of the forearm.

a.

b.

c.

17.3 Which of the muscles listed in 17.2 both extends and adducts the wrist?

17.4 Into which structures do the tendons of the extensor digitorum insert?

17.5 Name three muscles of the deep compartment that insert on the thumb.

a.

b.

c.

17.6 Which muscle of the deep compartment originates on the ulna and inserts on the index finger?

17.7 List the two bones of origin of the supinator muscle.

a.

b.

Exercise 18

Open the Cross Sections Book on the Library Shelf. Draw a cross section of the mid forearm using A.D.A.M. as a guide. Label all the muscles of the superficial and deep compartments on both the anterior and posterior forearm.

MUSCLES OF THE ANTERIOR AND MEDIAL THIGH

Exercise 19

The muscles of the anterior and medial thigh originate on the pelvis or the femur and cross the hip and/or knee joint to produce movements of the thigh and leg. Choose the Anterior view and review the muscles in the Tensor Fasciae Latae Muscle and the Vastus Medialis Muscle layers. Name each muscle described below.

19.1 originates on the iliac fossa and inserts on the femur ____________________

19.2 located on lateral side of upper thigh ____________________

19.3 crosses over thigh from lateral to medial to insert below the knee ____________________

19.4 only muscle of quadriceps femoris group to cross both hip and knee joints ____________________

19.5 largest adductor muscle of medial thigh ____________________

19.6 muscle of quadriceps femoris group that lies under rectus femoris ____________________

19.7 originates on pubis near pubic symphysis and inserts on femur near the adductor magnus ____________________

19.8 lateral muscle of the quadriceps femoris muscle group ____________________

MUSCLES OF THE POSTERIOR HIP AND THIGH

Exercise 20

The muscles of the posterior hip or gluteal region have several different actions on the femur. Choose the Posterior layer and look at the Gluteus Maximus Muscle, Gluteus Medius Muscle, and Gluteus Minimus Muscle layers. Describe the origin, insertion, and action of each of the three gluteal muscles.

Muscle	Origin	Insertion	Action
20.1 gluteus maximus			
20.2 gluteus medius			
20.3 gluteus minimus			

Exercise 21

The lateral rotators of the thigh at the hip can be identified in the Posterior view and the Piriformis Muscle, Quadratus Femoris Muscle, and Obturator Externus Muscle layers.

21.1 Between which two lateral rotators does the sciatic nerve emerge?

21.2 List the six lateral rotators of the thigh.

a.

b.

c.

d.

e.

f.

Exercise 22

The muscles of the posterior thigh extend the thigh and flex the leg because they cross both the hip joint and the knee joint. They are sometimes called the "hamstring" muscles, and include the biceps femoris, semitendinosus, and semimembranosus muscles. Review the Semitendinosus Muscle and Semimembranosus Muscle layers in the Posterior view.

22.1 Which of the posterior thigh muscles is the most lateral in position?

22.2 Which part of the pelvic bone is the origin of all three "hamstring" muscles?

22.3 Which of the medial "hamstring" muscles is superficial?

22.4 Between which two muscles of the posterior thigh does the sciatic nerve descend toward the leg?

MUSCLES OF THE ANTERIOR AND LATERAL LEG

Exercise 23

The muscles of the anterior compartment of the leg dorsiflex the ankle, and some of the muscles also extend the toes. Review these muscles by selecting the Anterior view and choosing the Tibialis Anterior muscle and the Extensor Digitorum Longus Muscle layers.

23.1 Describe the location and anatomy of the superior and inferior extensor retinacula.

a. superior extensor retinaculum

b. inferior extensor retinaculum

23.2 Which of the muscles of the anterior compartment lies adjacent to the tibia?

23.3 What is the insertion of the extensor digitorum?

23.4 Name the muscle that is continuous with the extensor digitorum and inserts on the fifth metatarsal.

23.5 Which muscle lies deep between the tibialis anterior and the extensor digitorum and acts to extend the big toe?

Exercise 24

The muscles of the lateral leg act to flex and evert the foot at the ankle. Select the Lateral view and the Peroneus Longus Muscle layer.

24.1 Name the two muscles of the lateral compartment.

a.

b.

24.2 What is the insertion of the peroneus brevis?

24.3 Name the connective tissue structures that keep these tendons in position on the lateral side of the ankle.

a.

b.

MUSCLES OF THE POSTERIOR LEG

Exercise 25

Plantar flexion of the foot at the ankle joint results from action of the muscles of the posterior leg. Some muscles also invert the foot and flex the toes. The muscles are arranged in superficial and deep compartments. Muscles of the superficial compartment can be seen in the Posterior view in the Gastrocnemius Muscle and Plantaris Muscle layers. The deep compartment muscles are seen in the Flexor Hallucis Longus Muscle and Tibialis Posterior Muscle layers. Answer T if the statement is true and F if the statement is false about the muscles of the superficial and deep compartments of the leg.

25.1 ____ The triangular muscle of the posterior knee is the tibialis posterior.

25.2 ____ Three muscle tendons of the superficial compartment insert into the calcaneus.

25.3 ____ The most lateral muscle in the deep compartment is the flexor digitorum longus.

25.4 ____ The tendon of the plantaris muscle descends toward the foot between the gastrocnemius and soleus muscles.

25.5 ____ The tendon of flexor digitorum longus crosses over the tendon of flexor hallucis longus at the ankle.

25.6 ____ The plantaris lies on the interosseous membrane between the flexor hallucis longus and flexor digitorum longus muscles in the deep compartment.

Exercise 26

Open the Cross Sections book on the Library Shelf. Draw a cross section of the mid leg and label all the muscles of the anterior, lateral, and posterior leg compartments. If you need help, use A.D.A.M. as a guide.

4

NERVOUS SYSTEM

STUDENT OBJECTIVES

Overview

- Review basic information on the anatomy and physiology of the nervous system.
- Name sensory receptors associated with the nervous system and explain their function.
- Describe the function of the major parts of the central nervous system.

The Brain and Cranial Nerves

- Identify the external features of the brain, including lobes, fissures, gyri, and sulci.
- Describe the location of the falx cerebri and the blood sinuses associated with it.
- Identify major internal structures of the brain, including the location of the components of the ventricular system.
- Draw and label a longitudinal section of the brain through the lateral ventricles.
- Identify several cranial nerves and some of their branches, which can be seen in lateral view.

The Spinal Cord and Spinal Nerves

- Describe the gross anatomy of the spinal cord and the meningeal coverings of the cord.
- Identify the components of a spinal nerve.
- Discuss the naming and numbering of spinal nerves based on their site of exit from the spinal cord.
- Describe the formation of the cervical, brachial, lumbar, and sacral plexuses, and identify many of the major nerves that arise from these plexuses.

The Autonomic Nervous System

- Identify the components of the sympathetic division of the autonomic nervous system in the thorax.
- Identify the major nerves and plexuses of the autonomic nervous system in the abdomen and pelvis.

OVERVIEW

Exercise 1

Click the View button in the Menu bar and select the Nervous System. Read the text overview and fill in the blanks in the following statements about the nervous system.

1.1 The central nervous system is composed of the ______________ and the ______________.

1.2 Another name for the cell body of a neuron is the ______________.

1.3 ______________ are chemicals that carry nerve signals across synapses between neurons.

1.4 Myelin sheaths that cover some axons are produced by the ____________ in the peripheral nervous system.

1.5 The hemispheres of the cerebrum communicate with each other by the ______________.

1.6 The largest part of the brain by weight is the ______________.

1.7 Shallow grooves on the surface of the brain that lie between gyri are ____________.

1.8 The three parts of the diencephalon are the ____________, ____________, and ______________.

1.9 ____________ are folds on the surface of the cerebellum.

1.10 The central constricted area of the cerebellum that connects its hemispheres is the ______________.

1.11 The most inferior portion of the brainstem is the ______________,

1.12 The network of nerves that extends throughout the midbrain and is essential for consciousness, awareness, and sleep is the ______________.

1.13 The ______________ is the tunnel in the vertebral column that transmits the spinal cord.

1.14 The layer of meninges that contains small blood vessels supplying the brain and spinal cord is the ______________.

1.15 The ______________ nervous system controls movements of skeletal muscles, whereas the ______________ nervous system controls movements of cardiac and smooth muscles.

1.16 The cell bodies of origin of the motor neurons that innervate skeletal muscles are located in the ______________ of the spinal cord.

1.17 The __________ root of the spinal nerve contains sensory fibers.

1.18 The division of the autonomic nervous sytem that responds during stress by increasing blood pressure and breathing rate is the ______________.

1.19 The ______________ division of the autonomic nervous system has a calming effect on the body and is responsible for promoting digestion and elimination.

1.20 The five necessary components of a reflex arc include ____________, ____________, ______________, ______________, and ______________.

Exercise 2

After reading the overview text, you should be able to name the structure or organ receptor described below.

2.1 ______________ receptors beneath the skin that sense pain

2.2 ______________ provide color vision

2.3 ______________ hearing receptor in the inner ear

2.4 ______________ senses deep pressure in the skin

2.5 ______________ apparatus in the inner ear that senses equilibrium

2.6 ______________ provide dim-light and black-and-white vision

2.7 ______________ detect changes in position of skeletal muscles

Exercise 3

Based on your reading of the text overview, complete the table below by describing the function of each part of the central nervous system.

Structure	Function
3.1 Cerebrum	
3.2 Hypothalamus	
3.3 Thalamus	
3.4 Epithalamus	
3.5 Cerebellum	
3.6 Midbrain	
3.7 Pons	
3.8 Medulla oblongata	
3.9 Spinal cord	

Exercise 4

Look at the image in the Primary window. Name as many of the structures of the Nervous System as you know. If you need help you can click the Label button. You can also listen to the correct pronunciation of the name by choosing that option from Supplements on the Menu bar.

THE BRAIN AND CRANIAL NERVES

Exercise 5

You are now ready to begin your review of the brain and cranial nerves. Click the View button on the Menu bar and select Full Anatomy. The external features of the lateral brain can be seen by selecting the Lateral view and the Brain layer from the Structure List window.

5.1 Identify the central sulcus. What are the names of the gyri that are located anterior and posterior to this sulcus?

5.2 Which other sulcus can be seen in this view?

5.3 Two of the lobes of the cerebrum are labeled in this image. What are they?

a.

b.

5.4 Identify and name two other lobes that are not labeled but can be seen in this view. You may need to consult your textbook for help if you do not know the answer.

a.

b.

Exercise 6

The central nervous system is surrounded by meninges (connective tissue coverings), which function to protect the brain and spinal cord, contain cerebrospinal fluid, form partitions between parts of the skull, and surround blood sinuses in the skull. One of these partitions (the falx cerebri), and its associated structures, can be seen when selecting the Lateral view and choosing the Falx Cerebri layer from the Structure List window.

6.1 Describe the location of the falx cerebri within the skull.

6.2 Which part of the ethmoid bone is the anterior attachment of the falx cerebri?

6.3 Name three midline blood sinuses that are formed within the falx cerebri.

a.

b.

c.

6.4 Where do the superior sagittal, straight, occipital, and transverse blood sinuses intersect?

Exercise 7

When you choose the Medial view and select the Central Nervous System and Sacral Plexus layer from the Structure List window, you can examine the internal structure of the brain.

Identify the following features or structures, and match them with the area of the brain with which they are associated.

		Structure	Area of Brain
7.1	_____	inferior colliculus	a. cerebrum
7.2	_____	interthalamic adhesion (intermediate mass)	b. medulla oblongata
			c. hypothalamus
7.3	_____	calcarine sulcus	d. midbrain
7.4	_____	inferior olivary nucleus	e. pons
7.5	_____	superior colliculus	f. thalamus
7.6	_____	mammillary body	
7.7	_____	corpus callosum	
7.8	_____	red nucleus	

Exercise 8

The ventricular system in the brain can also be reviewed using the Central Nervous System and Sacral Plexus layer in the Medial view. Name the region or regions of the brain that surround(s) each of the following spaces.

8.1 cerebral aqueduct

8.2 fourth ventricle

8.3 lateral ventricle

8.4 third ventricle

Exercise 9

Many of the structures you have seen in the lateral and medial views can also be seen in a horizontal (transverse) section of the brain. Open the Cross Sections book on the Library Shelf (Macintosh users only). Click the first button to show a horizontal section of the brain through the lateral ventricles, and review the structures you have previously identified in other views.

After you have completed your review, draw a sketch of a horizontal section of the brain through the lateral ventricles, using A.D.A.M. as a guide. Label in your drawing the following: frontal lobe, temporal lobe, occipital lobe, cerebral cortex, cerebral white matter, longitudinal fissure, lateral sulcus, falx cerebri, dura mater, corpus callosum, anterior and posterior horns of the lateral ventricles, choroid plexus, and thalamus.

Exercise 10

The twelve pairs of cranial nerves provide sensory and motor innervation to structures of the head and neck. Click on the Lateral view and select the Facial and Auriculotemporal Nerves layer from the Structure List window. The branches of one of the cranial nerves, the facial nerve, can be traced as they exit from the parotid gland on the face. Name these branches.

a.

b.

c.

d.

e.

Exercise 11

Several other cranial nerves can be seen in the Lateral view by selecting Cranial Nerves from the Structure List window.

11.1 List below the name and number of at least four other cranial nerves that you can see in this view.

a.

b.

c.

d.

11.2 Which of the cranial nerves has a branch to the carotid sinus?

11.3 Name the cranial nerve that continues to descend in the neck toward the thorax and abdomen.

11.4 What are the three divisions of the trigeminal nerve?

a.

b.

c.

THE SPINAL CORD AND SPINAL NERVES

Exercise 12

The meninges and gross anatomy of the spinal cord can be seen when you select the Posterior view and begin your examination using the Posterior Vertebral Column layer in the Structure List window. Scroll down the list, layer by layer, until you reach the Spinal Cord layer. Examine the anatomy of the spinal cord and meninges in each layer, and answer the questions below based on your review. You can also look at the Medial view and Central Nervous System and Sacral Plexus layer to help with your answers.

12.1 Where does the spinal cord begin?

12.2 Name the outermost meningeal covering of the cord.

12.3 Which meningeal covering lies deep to the covering described in 12.2?

12.4 The spinal cord is wider in two regions. Where are these enlargements located?

a.

b.

12.5 What is the conus medullaris, and where is it located in most adults?

12.6 What is the name of the thread of pia mater that anchors the tip of the spinal cord to the coccyx?

Exercise 13

Make sure you are looking at the Posterior view and the Spinal Cord layer, and complete the statements below about spinal nerves.

13.1 A spinal nerve is composed of ____________ and ____________ roots. (*Hint: Look at thoracic nerves T7, T8, and T9.*)

13.2 The collection of nerve roots within the dural sac and located inferior to the conus medullaris is the ________________.

13.3 The enlarged region of the dorsal root that contains sensory neuron cell bodies is the ________________.

13.4 Following the union of the spinal nerve roots to form a spinal nerve, the nerve splits into the ____________ ramus and the ____________ ramus.

13.5 Intercostal nerves are formed from the ____________ ramus of the spinal nerve.

13.6 The ____________ ramus of the spinal nerve innervates the muscles of the deep back.

13.7 ____________ rami in the thoracic region connect the spinal nerve with the autonomic nervous system.

13.8 The total number of pairs of spinal nerves that emerge from the spinal cord is ________.

13.9 There are ________ pairs of cervical nerves, ________ pairs of thoracic nerves, ________ pairs of lumbar nerves, ________ pairs of sacral nerves, and ________ pairs of coccygeal nerves.

13.10 Cervical nerve C6 exits the spinal cord ____________ (above or below) the sixth cervical vertebra.

13.11 Thoracic nerve T9 exits the spinal cord ____________ (above or below) the ninth thoracic vertebra.

13.12 Networks of spinal nerves in the cervical, lumbar, and sacral regions are called ____________.

Exercise 14

The cervical plexus is formed from the ventral rami of the first four cervical nerves in the neck region. You can see two of the cutaneous nerves that arise from this plexus by selecting the Anterior view and the Cutaneous Nerves layer from the Structure List window. Name the two cutaneous nerves that overlie the platysma muscle in this view of the neck.

a.

b.

Exercise 15

You can identify these two nerves described in Exercise 14 and others which have been cut by selecting the Inferior Root of Ansa Cervicalis layer from the Structure List window.

15.1 Name two other cutaneous nerves that arise from the cervical plexus that you can see in this layer.

a.

b.

15.2 Which two cervical nerves contribute to the formation of the inferior root of the ansa cervicalis?

a.

b.

15.3 Which cervical levels contribute to the formation of the phrenic nerve?

Exercise 16

The brachial plexus, which arises from ventral rami of spinal nerves C5-T1, supplies structures in the upper limb. Click the View button on the Menu bar, and select the Nervous System. Using this view as a guide, draw a rough sketch of the brachial plexus of nerves, beginning with its origin from cervical and thoracic nerves to its termination in the hand. On your drawing, label the roots, trunks, divisions, cords, and major terminal nerves (axillary, radial, musculocutaneous, median, and ulnar).

If you have trouble naming these structures, return to Full Anatomy on the Menu bar. Select the Anterior view, and review the Deep Nerve Plexuses and Intercostal Nerves layer and the Brachial Plexus and Branches layer in the Structure List window.

Exercise 17

Answer the following questions about the brachial plexus after you have reviewed the layers in Anterior view listed in Exercise 16 and the Radial Nerve layer, Median Nerve layer, Ulnar Nerve layer, and the Branch of the Ulnar Nerve layer from the Structure List window.

17.1 Which roots form the following?

a. upper trunk

b. middle trunk

c. lower trunk

17.2 The lateral, medial, and posterior cords are named because of their relationship to which artery?

17.3 The musculocutaneous nerve is a continuation of which cord?

17.4 Which muscle does the musculocutaneous nerve pass through to reach the arm?

17.5 The musculocutaneous nerve lies between two muscles of the arm that act to flex the forearm. What are they?

a.

b.

17.6 The median nerve is formed by contributions from which two cords?

a.

b.

17.7 How many branches does the median nerve give off in the arm region?

17.8 The median nerve accompanies which artery and vein through the elbow region?

17.9 Which cord continues to become the ulnar nerve?

17.10 The ulnar nerve in the forearm travels between which two muscles?

a.

b.

17.11 Most of the muscles in the palm of the hand are innervated by which terminal nerve of the brachial plexus?

17.12 Name the two terminal nerves of the posterior cord that supply muscles of the shoulder, posterior arm, forearm, and hand.

a.

b.

17.13 List five other nerves that arise from the brachial plexus to supply muscles in the shoulder and thorax region.

a.

b.

c.

d.

e.

Exercise 18

Ventral rami of spinal nerves L1-L4 form the lumbar plexus. Choose the Anterior view and the Deep Nerve Plexuses and Intercostal Nerves layer from the Structure List window. Using this view as a guide, draw a rough sketch of the lumbar plexus and label the following nerves: iliohypogastric, ilioinguinal, genitofemoral, lateral femoral cutaneous, femoral, and obturator.

Exercise 19

The sacral plexus is formed from ventral rami of L5-S3 with contributions from L4 and S4. Nerves from this plexus supply structures of the posterior hip and lower limb. Its branches to the hip can be seen by selecting the Posterior view and the Superior and Inferior Gluteal Nerves layer from the Structure List window. List five nerves arising from the sacral plexus that can be seen supplying structures in this region.

a.

b.

c.

d.

e.

Exercise 20

The large sciatic nerve exits the hip region to supply the lower limb. The course of this nerve can be traced as it descends toward the knee and foot by selecting the Posterior view and the Sciatic nerve and branches layer and the Anterior view and the Anterior Tibial Artery layer.

20.1 Which two muscles had to be removed in order for you to follow the sciatic nerve through the thigh? (You may have to scroll up the Stucture List window three or four layers in Posterior view to answer this question.)

a.

b.

20.2 Into which two major nerves does the sciatic nerve branch?

a.

b.

20.3 After the sciatic nerve branches, one of its major branches, listed in Exercise 20.2, divides again into which two nerves?

a.

b.

20.4 In general, which of the nerves described in Exercises 20.2 and 20.3 innervates muscles in each of the following compartments?

a. anterior compartment of leg muscles

b. lateral compartment of leg muscles

c. posterior compartment of leg muscles

THE AUTONOMIC NERVOUS SYSTEM

Exercise 21

The autonomic nervous system is composed of sympathetic and parasympathetic divisions. Some of the components of the sympathetic division can be seen by choosing the Anterior view and the Sympathetic Trunk layer from the Structure List window. Draw a rough sketch of the sympathetic chain, with its associated ganglia and its connections to the intercostal nerves in the thorax. Label the following structures in your drawing: sympathetic chain, chain ganglia, cervicothoracic ganglia, middle and superior cervical ganglia, thoracic splanchnic nerves, white and gray rami communicans, and intercostal nerves.

Exercise 22

Networks of both sympathetic and parasympathetic nerves travel through plexuses located deep in the abdomen and pelvis, which are closely associated with the arteries in the region. Some fibers synapse in the ganglia in the plexuses; others do not.

Select the Anterior view and the Autonomic Nerve Plexuses layer from the Structure List window. Draw and label the abdominal aorta and its major branches using A.D.A.M. as a guide. After you have drawn the artery, add the following structures to your drawing: vagal trunk, thoracic splanchnic nerves, celiac ganglia, superior mesenteric ganglion, aorticorenal ganglia, intermesenteric plexus, inferior mesenteric ganglion, superior hypogastric plexus, right and left hypogastric nerves, and the inferior hypogastric plexuses.

5

ENDOCRINE SYSTEM

STUDENT OBJECTIVES

Overview

- Review basic information on the anatomy and physiology of the endocrine system.
- Name the hormone(s) produced by each endocrine gland and describe their actions in the body.

Pituitary Gland

- Describe the location of the pituitary gland in relationship to other structures in the region.
- Name the parts of the pituitary gland and its attachment to the hypothalamus.

Thyroid Gland

- Describe the location of the thyroid gland in the neck.
- List the arterial supply and venous drainage of the thyroid gland.

Adrenal Glands

- Describe the location of the adrenal glands in the abdomen.
- List the arterial supply and venous drainage of the adrenal glands.

Pineal Gland

- Describe the location of the pineal gland in the brain.

The other endocrine glands, the pancreas, ovaries, testes, and thymus gland, will be reviewed in other chapters. The parathyroid glands cannot be seen using A.D.A.M.

OVERVIEW

Exercise 1

Click on View in the Menu bar and select the Endocrine System. Read the text overview and circle the correct answer to each of the following questions.

1.1 ___________ are chemical messengers released from endocrine glands.

a. enzymes
b. hormones
c. tissue factors
d. plasma factors

1.2 Which of the following homeostatic activities do secretions from endocrine glands control?

a. energy metabolism
b. water and electrolyte balance
c. lactation
d. a, b, and c

1.3 Which of the following glands is not an endocrine gland?

a. adrenal gland
b. thyroid gland
c. liver
d. ovary

1.4 The main neural control center for the endocrine system is the ___________.

a. cerebral cortex
b. midbrain
c. hypothalamus
d. thalamus

1.5 _____________ feedback decreases the deviation from an ideal normal value and is the type of feedback mechanism utilized for regulating secretion by many of the endocrine glands.

a. positive
b. negative
c. internal
d. external

1.6 The secretion of oxytocin is a good example of a ________ feedback mechanism.
  a. negative
  b. internal
  c. external
  d. positive

1.7 Which of the following glands contains both anterior and posterior lobes and is referred to as the "master gland" because of its regulation of numerous important homeostatic activities in the body?
  a. adrenal gland
  b. parathyroid gland
  c. thyroid gland
  d. pituitary gland

1.8 The two hormones that are actually produced by neuron cell bodies of the hypothalamus and then transported to the posterior pituitary for secretion are:
  a. ADH and oxytocin
  b. ACTH and oxytocin
  c. TSH and ADH
  d. ADH and PTH

1.9 Which of the endocrine glands has two lobes connected by an isthmus and is located in the neck near the larynx?
  a. parathyroid gland
  b. adrenal gland
  c. thymus gland
  d. thyroid gland

1.10 Endocrine glands contain ducts that are utilized to transport hormones to target organs.
  a. true
  b. false

Exercise 2

Fill in the following table with the missing information on the endocrine glands and their hormones and the actions of these hormones. You can complete this exercise after you have read the text overview for Exercise 1.

Endocrine Gland	Hormone	Action
parathyroid	2.1	2.2
anterior pituitary	2.3	stimulates adrenal cortex to secrete steroid hormones
2.4	TSH	2.5
ovary	2.6	required for formation of an ovum and preparation of the uterus for implantation
2.7	T3 and T4	2.8
adrenal cortex	2.9	increases sodium reabsorption from the blood by the kidneys
testes	2.10	required for sperm formation
2.11	growth hormone	2.12
2.13	epinephrine and norepinephrine	prepares the body for "fight or flight" responses (increases heart rate and sugar levels)
2.14	2.15	causes milk production by the breasts
2.16	2.17	decreases blood calcium

Exercise 3

Review the location of each of the endocrine glands in the image in the Primary window in both the male and female by selecting Options in the Menu bar and choosing Male or Female. Identify the following glands:

pituitary pancreas
thyroid ovaries
adrenals testes

After you have completed your overview of the endocrine system, you are ready to examine some of the glands in more detail. The glands you will review in this chapter are the pituitary, thyroid, adrenals, and pineal (body). The parathyroid glands are located on the posterior side of the thyroid gland; you will not be able to see them using A.D.A.M. The pancreas, a mixed gland that produces digestive enzymes and the hormones insulin and glucagon, will be reviewed in the chapter on the digestive system. The ovaries and testes will be covered in the reproductive system chapter. The thymus gland, which is largest in infants and children, will be examined with the lymphatic system.

PITUITARY GLAND

Exercise 4

Using View in the Menu bar, select Full Anatomy. Click the Medial view button and select the Central Nervous System and Sacral Plexus layer from the Structure List window. Review the relationships of the pituitary gland to the brain and the bones of the cranium, and answer the following questions.

4.1 To which part of the brain is the pituitary gland most closely related?

4.2 By which structure is the pituitary gland connected to the brain?

4.3 Name the two lobes of the pituitary gland.

a.

b.

4.4 The pituitary gland is surrounded by which part of the sphenoid bone?

4.5 Name the paranasal sinus that is closely associated with the pituitary gland.

THYROID GLAND

Exercise 5

As you know, endocrine glands must have good arterial supply and venous drainage in order to function properly and release their hormones into circulation. In this exercise you will review the location of the thyroid gland in the neck and the arteries which supply the gland by selecting the Anterior view and the Superior Thyroid Artery layer from the Structure List window.

5.1 Where does the isthmus of the thyroid gland cross the trachea anteriorly?

5.2 List two arteries that supply the thyroid gland, and name the artery from which each one originates.

a.

b.

Exercise 6

After reviewing the arterial supply to the thyroid gland, select the Internal Jugular Vein and Tributaries layer from the Structure List window. This view allows you to "add" a layer of veins to the preceding layer of structures.

6.1 List three veins that drain the thyroid gland, and state where each vein empties.

a.

b.

c.

ADRENAL GLANDS

Exercise 7

The adrenal glands are also referred to as the suprarenal glands. Each gland contains both an outer cortex and an inner medulla and produces several hormones. Choose the Anterior view and the Suprarenal Gland layer from the Structure List window. Answer the following questions after you have reviewed the location of the adrenal glands and their arterial supply and venous drainage.

7.1 Describe the location of the adrenal glands.

7.2 List three arteries which supply each adrenal gland.

a.

b.

c.

7.3 How does the venous drainage of the adrenal glands differ on the left and right sides of the body? (Where does each vessel drain?)

PINEAL GLAND

Exercise 8

The pineal gland, which secretes melatonin, is a small endocrine gland whose function is not completely known. Locate the gland in the brain using the Medial view and the Central Nervous System and Sacral Plexus layer from the Structure List window.

8.1 Describe the location of the pineal gland.

6

CARDIOVASCULAR SYSTEM

STUDENT OBJECTIVES

Overview

- Review basic information on the anatomy and physiology of the cardiovascular system.

The Heart

- Discuss the relationship of the pericardial sac to the heart and other structures in the region.
- Describe the external anatomy of the anterior heart and its associated great vessels.
- Name the arteries that supply the anterior heart wall, and the veins that drain the wall.
- Describe the internal anatomy of the heart, including its chambers and the location of the tricuspid valve and the pulmonary (semilunar) valve.
- Identify the chambers of the heart and its associated great vessels on a cross section taken through the thorax.

The Blood Vessels

- Identify the major vessels that form the pulmonary circuit. Describe the pathway of blood flow from the heart to the lungs and back to the heart.
- Identify and describe the location of the major arteries and veins of the systemic circuit in the head and neck, upper limbs, thorax, abdomen and pelvis, and lower limbs.
- Identify the three tributaries of the hepatic portal vein.

OVERVIEW

Exercise 1

Click on View in the Menu bar and select the Cardiovascular System. Read the text overview and circle the correct answer for the following questions, based on the text.

1.1 Which of the following is not a function of the cardiovascular system?

   a. transporting oxygen to body cells and removing cell waste
   b. transmitting electrical impulses to muscle cells
   c. protecting the body against microscopic foreign bodies
   d. helping in body heat regulation

1.2 Plasma, the liquid portion of blood, represents about ______ of total blood volume.

a. 35%
b. 45%
c. 55%
d. 65%

1.3 Which of the following cellular elements of the blood increase during an infection?

a. platelets
b. erythrocytes
c. sickle cells
d. leukocytes

1.4 If a patient is bleeding and not clotting properly, the cellular elements that may be decreased in the blood are the ________.

a. erythrocytes
b. platelets
c. leukocytes
d. white blood cells

1.5 Which of the following vessels are the smallest in diameter?

a. veins
b. arterioles
c. venules
d. capillaries

1.6 The epicardium of the heart wall is the ________.

a. visceral pericardium
b. fibrous pericardium
c. parietal pericardium
d. pericardial cavity

1.7 The two major vessels that deliver blood to the right atrium are the:

a. left and right pulmonary veins
b. left and right pulmonary arteries
c. aorta and superior vena cava
d. superior vena cava and inferior vena cava

1.8 The pulmonary ________ carry oxygen-poor blood from the heart to the lungs.

a. arteries
b. veins

1.9 Which of the following structures is the "pacemaker" of the heart?
a. AV bundle
b. SA node
c. left bundle branch
d. AV node

1.10 The amount of force that blood exerts on vessel walls (blood pressure) is affected by which of the following?
a. blood volume
b. peripheral resistance
c. cardiac output
d. a, b, and c

Exercise 2

Look at the image in the Primary window. See how many of the structures of the cardiovascular system you can identify without looking at the labels. After quizzing yourself, click the Label button and see how many structures you identified correctly. After you have identified a structure, listen to the correct pronunciation of the name of the structure and practice saying it.

THE HEART

Exercise 3

Using View in the Menu bar, select Full Anatomy. Examine the Anterior view and the Phrenic Nerve layer from the Structure List window. Identify the pericardial sac that surrounds the heart and highlight the sac with the Highlight option. After highlighting the sac, you may want to click the Zoom button and select the Zoom In option in the Primary window to see the structures more clearly.

3.1 To which structure does the pericardial sac attach inferiorly?

3.2 Name the nerve and blood vessels that have a very close relationship to the pericardial sac.

a. _____________ nerve

b. _____________ artery

c. _____________ vein

Exercise 4

Now you are going to perform surgery and open the pericardial sac to expose the heart in Anterior view. Open the OR Panel by clicking the OR button in the Primary window. Click and hold the Scalpel button and select Wide for the incision width. Make an incision from the upper right side of the heart diagonally toward the lower left side of the heart (the apex of the heart). Make another incision parallel to the initial incision, but more inferior, so that the entire heart is exposed.

4.1 What do you see when the sac is opened?

4.2 What is the fluid that fills the space between the pericardial sac and the heart wall? (*Hint: You read about it in the text overview.*)

4.3 What is the yellow material you see covering the vessels on the anterior of the heart?

Exercise 5

Leave the OR and return to the Structure List window and select the Cardiac Veins layer in Anterior view. We will examine this layer to identify the chambers and vessels of the heart. In order to see these structures more clearly, make sure that you are in the highest magnification possible, using the Zoom button. Based on your observations, fill in the blanks with the correct answers in the following paragraphs. Use the Identify button to help you find the answers.

Three of the four chambers of the heart can be seen in this anterior view. These are the (5.1) ____________________, (5.2) ____________________, and (5.3) ____________________. The (5.4) ____________________ of the left atrium is also slightly visible. The large vein that empties into the right atrium is the (5.5) ____________________. The large blue structure that exits from the right ventricle is the (5.6) ____________________. It branches into the (5.7) ____________________ and (5.8) ____________________ arteries. The left branch immediately enters the (5.9) ____________________. The right branch travels toward the right lung (5.10) ____________________ (anterior or posterior) to the other large vessels. The large red vessel is the (5.11) ____________________.

The coronary vessels supply and drain the heart wall. The coronary artery, which travels between the right atrium and right ventricle and has branches that supply the two chambers, is the (5.12) ____________________. The (5.13) ____________________ branch of this artery courses along the inferior margin of the heart toward the apex. This artery, described in 5.13, is accompanied by the (5.14) ____________________ vein. Other veins that drain directly into the right atrium on its anterior surface are the (5.15) ____________________ veins. The other major coronary artery, the left coronary artery, cannot be seen in this view. The branch of the left coronary artery that supplies the left and right ventricles is the (5.16) ____________________ . The vein that drains the ventricles and accompanies the artery is the (5.17) ____________________ .

Exercise 6

While still in Zoom In and Anterior view, scroll down the Structure List window and choose the Heart layer. All of the coronary vessels have been removed to show the depressions on the surface of the heart where the vessels are located. Identify the depressions and name the structures located in each depression. You may want to go back and forth between the Cardiac Veins layer and the Heart layer to help you answer this question.

	Depression	Structure(s)
6.1		
6.2		

Exercise 7

Scroll down the Structure List window and select the Heart—Cut Section layer. Identify the three chambers of the heart you listed in Exercise 5, which are now visible in the sectioned heart.

7.1 Name the depression in the wall of the right atrium.

7.2 Which chamber has the thickest muscular wall?

7.3 Why do you think it is thicker than the other walls?

Exercise 8

Valves direct the flow of blood through the heart and are located between the atria and ventricles and between the ventricles and the large arteries as they exit the chambers. Select the Heart—Cut Section layer from the Structure List window. Two of these valves can be seen in this view.

8.1 Name the two valves and describe their location in the heart.

a.

b.

8.2 Do the two valves have the same structure?

8.3 Identify the papillary muscles in the ventricles. What are the white cord-like structures that attach between the papillary muscle and the undersurface of the tricuspid valve?

8.4 What is their function?

Exercise 9

Scroll down the Structure List window and choose the Pericardial Sac layer. The heart has been removed to allow viewing of the openings of the other vessels that empty into the heart.

9.1 Name the five vessels—other than the superior vena cava, aortic arch, and pulmonary trunk—whose openings can now be seen.

a.

b.

c.

d.

e.

9.2 Four of the openings you have just listed in 9.1 return oxygenated blood to the heart as part of the pulmonary circuit. Where are they delivering the blood from?

9.3 Refer to Exercise 5 and 9.2 and name the vessels, in order, that transport blood from the heart to the lungs and back.

Exercise 10

Now that you have reviewed the external and internal heart in Anterior view, open the Cross Sections book on the Library Shelf and view the heart in cross section. Click on the Cross Section button (sixth level from the top). Draw a rough sketch of the cross section and label the following structures to help you review the relationship of these structures with other structures in the thorax.

right atrium	superior vena cava
right ventricle	anterior cardiac vein
left ventricle	right and left coronary arteries
left atrium	great cardiac vein
aorta	phrenic nerves
pulmonary vein	

THE BLOOD VESSELS

The Aorta and its Branches

Exercise 11

The aorta is the largest artery in the body and carries blood away from the heart in the systemic pathway of blood circulation. Make sure you are in Full Anatomy and Anterior view, and select the Heart layer in the Structure List window. Examine the branches of this large vessel. You may want to Zoom In to enlarge the region.

11.1 What are the three major separate branches of the aortic arch?

a.

b.

c.

11.2 Into which two branches does the brachiocephalic trunk divide?

a.

b.

11.3 Which large vein lies anterior to these vessels?

11.4 Which nerve crosses the arch on its left side?

11.5 Where is this nerve's counterpart located on the right side?

Arteries of the Head and Neck

Exercise 12

The arteries that supply many of the structures of the head and neck arise from the common carotid artery. The left common carotid artery originates from the aortic arch, but the right common carotid artery comes from the brachiocephalic trunk. Examine the common carotid artery and its branches by selecting the Lateral view and the Common Carotid Artery and Branches layer from the Structure List window. You may want to Highlight the arteries in order to see their branches more clearly.

12.1 Name the two branches of the common carotid artery.

a.

b.

12.2 At what level in the neck does the common carotid artery divide?

12.3 List six of the branches of the external carotid artery that you can see in this view.

a.

b.

c.

d.

e.

f.

Arteries of the Brain

Exercise 13

The internal carotid and vertebral arteries furnish oxygenated blood to the brain. After the internal carotid artery branches from the common carotid artery, it ascends into the skull through the carotid canal to the brain. Its branches include the middle and anterior cerebral arteries. The anterior cerebral arteries and the vertebral arteries can be seen when you select the Aortic Arch and Branches layer from the Structure List window and the Lateral view.

13.1 Identify the left anterior cerebral artery in the brain. The artery lies very close to which structure of the cerebrum?

13.2 The vertebral artery can also be seen using this same view and layer. From which vessel in the neck does the vertebral artery arise?

13.3 The vertebral artery courses through the neck within the transverse foramina of the cervical vertebrae and into the skull through the foramen magnum with the spinal cord. Both vertebral arteries join to form a single vessel on the ventral surface of the pons. Which vessel is it?

13.4 Name three branches of the vessel described in 13.3.

a.

b.

c.

13.5 The artery described in 13.3 terminates by dividing into the left and right posterior ______________ arteries.

The course of several other arterial branches that supply the brain cannot be seen in A.D.A.M. and will be omitted in this chapter.

Arteries of the Upper Limbs

Exercise 14

The subclavian arteries carry blood toward the upper limbs. The left subclavian artery is a direct branch from the aorta and the right is a branch of the brachiocephalic artery. Return to the Anterior view in the Primary window and select the Axillary Artery and Branches, from the Structure List window. You may also want to use the Zoom and Highlight buttons.

14.1 Follow the course of the subclavian artery and name four of its branches.

a.

b.

c.

d.

14.2 Where does the subclavian artery become the axillary artery?

14.3 Name the branch of the axillary artery that supplies the lateral chest wall.

14.4 Two branches of the axillary artery furnish blood to the humerus. What are they?

a

b.

14.5 The axillary artery continues into the arm, changing its name to the ____________ artery.

14.6 What is the first major branch of the artery described in 14.5 after it enters the arm?

14.7 Name two other branches of the artery described in 14.5 that travel medially toward the elbow region.

a.

b.

14.8 The brachial artery crosses anterior to which muscle tendon in the elbow region?

14.9 Near the elbow, the brachial artery divides into which two arteries?

a.

b.

14.10 Which of the arteries listed in 14.9 continues *medially* into the hand to contribute to the formation of the superficial palmar arterial arch?

Branches of the Thoracic and Abdominal Aorta

Exercise 15

The continuation of the aortic arch into the thorax is the thoracic portion of the descending aorta. The descending aorta continues inferiorly into the abdomen as the abdominal aorta. While in Anterior view, select the Abdominal Aorta and Branches layer using the Structure List window. Answer T for true and F for false to the following questions about the thoracic and abdominal aorta.

15.1 ______ The aorta descends anterior to the pulmonary trunk and arteries in the thorax.

15.2 ______ Autonomic nerves lie very close to the aorta as it descends through the thorax.

15.3 ______ The opening in the diaphragm for the esophagus is inferior to the opening for the aorta.

15.4 ______ The first pair of arteries that arise from the aorta as it exits through the diaphragm are the inferior phrenic arteries.

15.5 ______ The most superior unpaired branch of the aorta is the celiac trunk.

15.6 ______ The paired superior mesenteric arteries supply the kidneys.

15.7 ______ The left gonadal (testicular or ovarian) artery arises more superior in position from the aorta than the origin of the right gonadal artery.

15.8 ______ The right common iliac artery lies posterior to the right common iliac vein.

15.9 ______ The external iliac artery has a close relationship to the psoas major muscle.

15.10 ______ The internal iliac artery continues into the thigh as the femoral artery.

Arteries of the Lower Limbs

Exercise 16

You can trace the proximal part of the femoral artery as it descends through the thigh using the same view and layer as you used in Exercise 15.

16.1 What is the largest branch of the femoral artery in the thigh?

16.2 Name the branch of the deep femoral artery that travels laterally to supply the head and neck of the femur.

Exercise 17

The femoral artery travels from the anterior thigh to the posterior knee region through a gap in the adductor magnus muscle called the adductor hiatus. The course through the hiatus can be seen when you choose the Medial view and select the Nerves of the Adductor Canal layer from the Structure List window.

17.1 What does the femoral artery become when it exits through the hiatus?

Exercise 18

Locate the popliteal artery in the posterior knee region by changing to the Posterior view and selecting the Posterior Tibial Artery layer from the Structure List window.

18.1 Which nerve and vein accompany the popliteal artery in the posterior knee region?

18.2 Identify and name the branches of the popliteal artery that supply the popliteal region of the knee.

18.3 The popliteal artery branches into the ________________ and ______________ arteries.

Exercise 19

The posterior tibial artery is one of the two major branches of the popliteal artery. Trace its course inferiorly through the posterior leg toward the foot, using the same view and layer as you used for Exercise 18.

19.1 Name three deep muscles of the leg that lie close to the artery as it descends through the leg toward the foot.

a.

b.

c.

19.2 What are the two terminal branches of the posterior tibial artery in the foot?

a.

b.

19.3 Identify the plantar digital arteries and describe their distribution.

Exercise 20

The largest branch of the posterior tibial artery is the peroneal artery. Its location can be seen when you select the Popliteal Artery and Branches layer from the Structure List window.

20.1 The peroneal artery lies close to which bone?

20.2 Does it cross the ankle medially or laterally?

Exercise 21

The other major branch of the popliteal artery is the anterior tibial artery. In order to trace its course, change to the Anterior view in the Primary window and select the Anterior Tibial Artery layer in the Structure List window. Use the Zoom button if you need to see more detail.

21.1 The anterior tibial artery emerges through an opening in the interosseous membrane between the tibia and the fibula. Which branch of the anterior tibial artery ascends toward the knee region soon after the anterior tibial artery emerges through the opening?

21.2 The anterior tibial artery descends in the leg with which nerve?

21.3 What does the anterior tibial artery become after it crosses the ankle region?

21.4 Name the branches of the dorsalis pedis artery that supply the tarsal bones.

a.

b.

21.5 What is the total number of dorsal metatarsal arteries in each foot?

21.6 How many digital arteries arise from each dorsal metatarsal artery?

Veins of the Head and Neck

Exercise 22

Begin your examination of the veins that drain structures in the head and neck by selecting the Lateral view and the External Jugular Vein and Tributaries layer from the Structure List window. The external jugular vein and its tributaries are relatively superficial in position and must be removed in order to see the deeper veins of the neck. Review the course of the internal jugular vein and its tributaries by "dissecting" deeper into the neck, using the scrolling bar and selecting the Tributaries of the Right Brachiocephalic Vein layer from the Structure List window. Answer the following questions about the veins in the neck after you have reviewed both layers.

22.1 Which muscle lies deep to the external jugular vein?

22.2 Into which vessel does the external jugular vein empty?

22.3 Name three tributaries of the internal jugular vein.

a.

b.

c.

22.4 The internal jugular vein and the subclavian vein join to form which vein?

22.5 Which cranial nerve travels with the internal jugular vein in the neck?

Veins of the Upper Limbs

Exercise 23

The veins that drain the upper and lower limbs are both superficial and deep in position. The superficial veins lie nearer the surface of the body and have connections to the deeper veins. We will begin our review of the veins of the upper limbs by choosing the Anterior view and Superficial Veins layer from the Structure List window.

23.1 What are the two large superficial veins of the arm and forearm?

a.

b.

23.2 They are connected at the elbow region by the ________________ vein.

Exercise 24

The deep veins in the upper limb drain blood toward the heart and travel with the arteries. Review the veins in the Brachiocephalic Vein and Tributaries layer in the Structure List window.

24.1 Name all of the deep veins, in order, that drain blood from the hand to the brachiocephalic vein. (*Hint: If you know the names of the arteries, you should know the names of the veins.*)

24.2 How many veins accompany each artery in the forearm?

Veins of the Thorax

Exercise 25

In Anterior view, select the Azygos Vein and Tributaries layer from the Structure List window. After reviewing the structures shown in the layer, complete the following statements about the azygos system of veins in the thorax.

25.1 The largest vein in the azygos system of veins that lies to the right of the vertebral bodies is the ________________ vein.

25.2 Two veins on the left posterior wall that empty into the azygos vein are the (a) ________________ and the (b) ________________.

25.3 The ________________ veins, which drain the chest wall, empty into the veins of the azygos system.

25.4 The azygos vein receives blood from the ________________ vein in the lumbar region.

25.5 The ________________ veins empty into the ascending lumbar veins.

Veins of the Abdomen and Pelvis

Exercise 26

The abdominal wall and abdominopelvic viscera are drained by veins that empty into the inferior vena cava. The tributaries that drain into the inferior vena cava can be seen when you select the Anterior view and the Inferior Vena Cava layer in the Structure List window.

26.1 The inferior vena cava is formed by the junction of which two veins?

26.2 Name two other pairs of veins that empty directly into the inferior vena cava.

a.

b.

26.3 Which is longer, the right or left renal vein?

26.4. In the male, do both testicular veins drain into the inferior vena cava? If not, where do they drain?

26.5 Which other vein is a tributary of the left renal vein?

26.6 The internal iliac vein drains structures in the pelvic and gluteal regions. Which vessel does it join to form the common iliac vein?

Hepatic Portal Circulation

Exercise 27

The portal vein and its tributaries transport venous blood from the digestive viscera to the liver. These can be seen in Anterior view by selecting the Portal Vein and Tributaries layer in the Structure List window.

27.1 Identify and name the three tributaries of the hepatic portal vein.

a.

b.

c.

Veins of the Lower Limbs

Exercise 28

The veins draining the lower limbs are also superficial and deep in position. The superficial veins in the lower limbs can be seen when you choose the Anterior view and the Superficial Veins layer in the Structure List window.

28.1 Which veins drain deoxygenated blood from the toes toward the dorsal venous arch?

28.2 On the lateral side of the foot, the dorsal venous arch empties into the ____________ vein.

28.3 Medially, the dorsal venous arch drains into the ____________________ vein.

Exercise 29

The deep veins of the lower limb can be reviewed using several different layers in the Structure List window in both the Anterior and Posterior views. Look at the veins in the following views and layers.

Posterior view	Posterior Tibial Vein layer
Posterior view	Femoral Vein and Tributaries layer
Anterior view	Anterior Tibial Vein layer
Anterior view	Inferior Vena Cava layer

29.1 Name the major veins, in order, that transport deoxygenated blood from the plantar surface of the foot to the external iliac vein.

29.2 Name the vein that drains the lateral side of the leg and empties into the posterior tibial vein.

29.3 Which vein returns blood from the dorsum of the foot to the popliteal vein?

7

LYMPHATIC SYSTEM

STUDENT OBJECTIVES

Overview

- Review basic information on the anatomy and physiology of the lymphatic system.

Thoracic Duct

- Describe the location of the thoracic duct in the abdomen and thorax.
- Name the parts of the body drained by the thoracic duct.

Lymphoid Organs

- Describe the location of major groups of lymph nodes.
- Describe the location, arterial supply, and venous drainage of the spleen.
- Describe the location of the thymus in the thorax.
- Describe the location of the palatine and lingual tonsils in the mouth.

OVERVIEW

Exercise 1

Click on View in the Menu bar, and select the Lymphatic System. Read the text overview and answer the following questions about the lymphatic system by marking T if the statement is true; F if the statement is false.

1.1 ______ Lymphatic vessels play an important role in maintaining tissue fluid balance by reabsorbing lymph, which bathes tissue cells, and then returning it to the circulation.

1.2 ______ Capillary hydrostatic pressure, sometimes called filtration pressure, forces fluid out of the blood capillaries into the surrounding tissues.

1.3 ______ The lymphatic system consists of a series of large tubes, which have tightly packed cells that form their walls, and that circulate lymph fluid similarly to the circulation of blood.

1.4 ______ Valves are located in lymph vessels and prevent the backward flow of lymph fluid.

1.5 ______ Lymph fluid from the right upper extremity is collected in the thoracic duct, which empties into the right subclavian vein.

1.6 ______ The origin of the thoracic duct, which transports lymph and fats from the digestive tract in the abdomen, is the cisterna chyli.

1.7 ______ The structures located along the routes of the lymphatic vessels that filter out foreign substances are called lymph nodes.

1.8 ______ Lymphoid structures include the spleen, thymus, tonsils, and appendix.

1.9 ______ Red pulp, which contains erythrocytes and macrophages, and white pulp, which contains lymphocytes, are types of tissue found in the thymus.

1.10 ______ Palatine, pharyngeal, and lingual tonsils are named because of their location in the oral cavity.

THORACIC DUCT

Exercise 2

The location of the thoracic duct in the abdomen and thorax can be reviewed by clicking View in the Menu bar and selecting Full Anatomy. Choose the Anterior view and the Thoracic Duct layer from the Structure List window.

2.1 Describe the relationships of the thoracic duct to the following structures as it ascends from the cisterna chyli to empty into the left subclavian vein.

a. vertebral column

b. azygos vein

c. hemiazygos vein

d. accessory hemiazygos vein

e. left sympathetic trunk

2.2 From which part(s) of the body does the thoracic duct collect lymphatic fluid?

LYMPHOID ORGANS

Lymphoid organs are collections of lymphoid tissue that vary in size and structure. The lymph nodes, spleen, thymus, and tonsils are examples of lymphoid organs, and will be reviewed in this chapter.

Lymph Nodes

Exercise 3

Lymph nodes are collections of lymphatic tissue that act as filter units along the lymphatic vessels of the body. They are also important in the activation of the immune response. Groups of lymph nodes are located in many places in the body and are generally named for their location. Return to View in the Menu bar and select the Lymphatic System. Review the location of the groups of lymph nodes in the image in the Primary window. Draw a simple sketch of the body and label the following lymphatic structures: thoracic duct, cisterna chyli, and facial, cervical, axillary, tracheal, bronchial, aortic, and inguinal lymph nodes.

Spleen

Exercise 4

The spleen is the largest lymphoid organ, and has many immunologic and hematologic functions. Click View in the Menu bar and choose Full Anatomy. Examine the Pancreas and Spleen layers in the Structure List window in the Anterior view. Answer the following questions based on your observations of the structures that you can see in the two layers.

4.1 Which organ leaves an "impression" on the surface of the spleen?

4.2 Name the large respiratory muscle that curves around much of the spleen on its superior and posterior borders.

4.3 The spleen is located near the "tail" of which organ?

4.4 Which major vessel that supplies the spleen with arterial blood lies posterior to the organ described in 4.3?

4.5 Name the major vein that drains deoxygenated blood from the spleen.

4.6 Which large organ lies near the inferior part of the spleen?

Thymus

Exercise 5

The thymus varies in size with the age of the individual. It is prominent in infants, continues to grow until adolescence, and then gradually atrophies. The thymus plays an important role in the development of T lymphocytes. Select the Anterior view and examine the anatomy of the thymus gland and its relationships to surrounding structures in the Ribs layer and the Remnant of Thymus Gland layer. Complete the following statements by writing the missing terms in the blanks.

The thymus lies posterior to the (5.1) ____________________ part of the sternum. Its superior border touches the (5.2) ____________________, which contains cartilaginous rings. The gland appears to "drape" over the large (5.3) ____________________ vein as the gland descends in the thorax. Two large arteries that are located adjacent to the thymus gland are the (5.4) ____________________ and the (5.5) ____________________ arteries. The

(5.6) ____________________, which surrounds the heart, lies posterior to the thymus gland.

Tonsils

Exercise 6

The tonsils are collections of lymphatic tissue around the entrance to the pharynx. They include the palatine, lingual, and pharyngeal tonsils. Select the Palatine Gland and Tonsils layer from the Structure List window in the Anterior view, and review the relationships of the palatine tonsils to other structures in the mouth.

6.1 The palatine tonsils are located between which two muscles of the posterior oral cavity in this view?

a.

b.

6.2 The lingual tonsil can be seen when you change to the Medial view and select the Tongue layer. Where is it located in the mouth?

8

RESPIRATORY SYSTEM

STUDENT OBJECTIVES

Overview

- Review basic information on the anatomy and physiology of the respiratory system.
- Describe the pathway along which air travels through the respiratory system during inspiration by listing all structures that transport air from the nasal cavity to the alveoli of the lung.

Pathway of Air Through the Respiratory System

- Describe the structures that form the external nose.
- Identify the four paranasal sinuses and describe their relationships to the nasal cavity.
- Name the bony structures that form the roof and floor of the nasal cavity and describe the bony projections from the lateral walls.
- Differentiate between the nasopharynx, oropharynx, and laryngopharynx.
- Name the three unpaired cartilages of the larynx.
- Identify some of the supporting structures that help to anchor the cartilages of the larynx to the hyoid bone and to each other.
- Describe the location of the two pairs of mucosal folds in the larynx.
- Describe the branching of the trachea into primary bronchi.
- Discuss the gross anatomy of the lung, including its lobes and fissures.
- Describe the relationships of the parietal and visceral pleura to the lungs and surrounding structures.

OVERVIEW

Exercise 1

Select View in the Menu bar, choose Respiratory System, and read the text overview. Circle the correct answer to the following questions.

1.1 Functions of the respiratory system include:

a. oxygenation of blood

b. maintenance of blood pH

c. regulation of body temperature

d. a, b, and c

1.2 Internal respiration involves oxygen and carbon dioxide exchange at the ____________ level.

a. system
b. cellular
c. organ
d. organism

1.3 In the lungs, oxygen and carbon dioxide exchange occurs across the ________________.

a. primary bronchi
b. secondary bronchi
c. alveolar sacs
d. tertiary bronchi

1.4 Humans breathe about ________ times per minute at rest.

a. 12
b. 18
c. 25
d. 30

1.5 Which structures in the nasal cavity push mucous containing foreign matter toward the pharynx?

a. thick nasal hairs
b. tiny cilia lining the nasal epithelium
c. microvilli of the nasal epithelium
d. muscle action of the trachea

1.6 The ________________ prevents food and liquids from entering the larynx during swallowing.

a. epiglottis
b. trachea
c. nasal concha
d. pharynx

1.7 The structure of the respiratory system that contains the vocal cords is the ________________.

a. nasal cavity
b. trachea
c. larynx
d. pharynx

1.8 What effect does contracting the diaphragm have on the volume of the thoracic cavity?

a. no effect
b. increases the volume
c. decreases the volume

1.9 What results from the action described in 1.8?

a. inspiration
b. expiration
c. ventilation
d. gas exchange

1.10 The fluid that allows the lungs to move freely in the pleural sac is located in the ________________.

a. pleural membrane
b. parietal pleura
c. visceral pleura
d. pleural cavity

PATHWAY OF AIR THROUGH THE RESPIRATORY SYSTEM

Exercise 2

After reading the text overview, you should be able to describe the pathway of air during the processes of inspiration and expiration. List, in order, all of the structures of the respiratory system that transport air from the nasal cavity to the alveoli in the lung.

External Nose

Exercise 3

The external nose is composed of bone, cartilage and dense connective tissue. Go to Full Anatomy by clicking the View button in the Menu bar. Choose the Anterior view and select the Nasal Cartilage layer in the Structure List window. You may want to enlarge the image by clicking the Zoom button and selecting Zoom In. Draw a simple sketch of the components of the external nose in the space provided below. Include in your drawing the following structures:

nasal bones	fat
lateral nasal cartilages	nasal septal cartilages
alar cartilages	maxilla

Paranasal Sinuses

Exercise 4

The nose is surrounded by the paranasal sinuses, which are spaces that help to warm and moisten the air. They are named for the bones in which they are located. One of these, the maxillary sinus, can be seen in Anterior view when you scroll down the Structure List window to the Tongue—Coronal Section layer.

4.1 Locate the maxillary sinus. Into which meatus of the internal nose does the maxillary sinus open?

Exercise 5

Change to the Lateral view and select the Trachea layer from the Structure List window. You can now identify three other sinuses that are closely associated with the nasal cavity.

5.1 What are they?

a.

b.

c.

Nasal Cavity and Pharynx

Exercise 6

The nasal cavity is a bony chamber lined by mucous membrane. It receives air through the external nares and is continuous posteriorly with the pharynx. The anatomy of the nasal cavity and the pharynx can be examined in the Medial view and the Nasal Conchae layer in the Structure List window. Fill in the blanks to complete the following paragraphs about the nasal cavity and pharynx.

The area of the nasal cavity located just inside the openings into the nose is the (6.1) _______________. The (6.2) ___________________ is part of the ethmoid bone, which forms the roof of the nasal cavity. The floor of the nasal cavity is formed by the palate, which consists of hard and soft portions. The two bones that contribute to the formation of the hard palate are the (6.3) ______________ and the (6.4) _________________. The soft palate is muscular. The hard and soft portions of the palate separate the nasal cavity from the (6.5) ______________ cavity. The three bony shelves that project from the lateral walls of the nasal cavity are the (6.6) _______________, (6.7)_______________, and (6.8) _________________ nasal conchae.

The pharynx is a muscular tube with three divisions. The most superior division, located posterior to the nasal cavity, is the (6.9) _________________. The (6.10) _______________ lies posterior to the oral cavity. The (6.11) __________________ is the most inferior division of the pharynx, and lies behind the larynx. The torus tubarius surrounds the opening of the (6.12) __________________, which connects the throat with the middle ear. The (6.13) __________________ tonsil is located in the roof of the nasopharynx.

Larynx

Exercise 7

The supporting structure of the larynx is primarily cartilaginous, rather than muscular. Nine separate cartilages, three paired and three unpaired, form the walls of the larynx. The three unpaired cartilages can be identified in the Anterior view and the Hyoid Bone and Thyroid Cartilage layer in the Structure List window. If you need to enlarge the image, use the Zoom button and choose Zoom In.

7.1 Two of these cartilages are anterior in position. Which ones?

a.

b.

7.2 The third unpaired cartilage is posterior to the hyoid bone. Name it.

7.3 What is the supporting membrane that attaches the thyroid cartilage to the hyoid bone?

7.4 Name the ligament that lies between the thyroid and cricoid cartilages.

7.5 What is the muscle that lies close to the ligament described in 7.4?

Exercise 8

The internal anatomy of the larynx can be seen in the Medial view and the Nasal Conchae layer in the Structure List window.

8.1 Name the two pairs of horizontal mucosal folds that project from the internal walls of the larynx.

a.

b.

8.2 Which pair of folds in 8.1 is more superior in position?

8.3 Which folds are the true vocal cords?

Trachea and Bronchi

Exercise 9

Click on Anterior view and select the Trachea layer in the Structure List window to review the anatomy of the trachea and bronchi. The trachea is the continuation of the larynx that conducts air toward the lungs. It is formed by 16-20 cartilaginous rings, which are incomplete on its posterior surface. The trachea terminates by branching into two smaller conducting tubes.

9.1 What are the names of the distal branches of the trachea?

9.2 Are the two branches the same size? If not, how are they different?

9.3 Do they branch from the trachea at the same angle? Support your response with anatomical descriptions.

Lungs and Pleural Coverings

Exercise 10

The gross anatomy of the lungs and the pleural coverings can be reviewed by selecting the Anterior view and choosing the Lungs layer in the Structure List window. Answer T for true and F for false to the following statements, based on your examination of the structures in the Lungs layer.

10.1 ______ The parietal pleura covers the superior diaphragm.

10.2 ______ The left lung is larger than the right lung.

10.3 ______ The superior tip of the lung is called the base.

10.4 ______ The cardiac notch is located on the right lung, and is the impression on the surface of the lung for the heart.

10.5 ______ Each lung is divided into three lobes.

10.6 ______ The oblique fissure on the left lung subdivides the lung into superior and middle lobes.

10.7 ______ The middle lobe of the right lung lies between the horizontal and oblique fissures.

Exercise 11

Open the Cross Sections book on the Library Shelf and view the lungs in cross section. Click on the Cross Section button (sixth level from the top). This section will allow you to review the relationships of the parietal and visceral pleura to each other and to surrounding structures.

Answer the following questions based on your examination of the cross section.

11.1 Which of the layers of the pleural coverings overlies the surface of the lung?

11.2 What is the relationship of the visceral pleura to the fissures on the surface of the lung?

11.3 What is the relationship of the parietal pleura to the ribs and vertebrae?

9

DIGESTIVE SYSTEM

STUDENT OBJECTIVES

Overview

- Review basic information on the anatomy and physiology of the digestive system.
- Name the components of the digestive tract in order, beginning with the mouth and ending with the anus.

Oral Cavity and Pharynx

- Describe the muscular attachment of the tongue and identify one of the intrinsic muscles that forms this muscular structure.
- Name the superior boundary of the oropharynx and the digestive system continuation of the inferior boundary of the laryngopharynx.

Teeth

- Describe the location of different types of teeth, including incisors, canines, premolars, and molars.

Salivary Glands

- Describe the location of the three pairs of salivary glands.
- Describe the course of the parotid duct from the gland into the mouth.
- Identify important nerves that have close relationships to the glands.

Esophagus, Stomach, Small Intestine, and Large Intestine

- Describe the location and gross anatomy of the esophagus, stomach, small intestine, and large intestine.
- Describe the relationship of the anterior vagal trunk to the esophagus.

Liver, Gall Bladder, and Pancreas

- Describe the location and gross anatomy of the liver, gall bladder, and pancreas.
- Name the major vessels that enter and exit the liver.
- Name the ducts that transport bile from the liver and gall bladder to the duodenum.

Peritoneum

- Define the following terms: parietal and visceral peritoneum; peritoneal cavity; retroperitoneal and peritoneal reflections.
- Describe the attachments of the greater omentum, lesser omentum, falciform ligament, transverse mesocolon, and mesentery.

OVERVIEW

Exercise 1

Click on View in the Menu bar and select Digestive System. Read the text overview and answer the following questions about the digestive system by marking T if the statement is true or F if the statement is false. If the statement is false, rewrite the statement by striking out the incorrect word(s) and inserting the correct word(s).

1.1 _____ Activities that occur during the digestive process include mechanical and chemical breakdown of food stuffs, transportation of ingested material, absorption, and defecation.

1.2 _____ The enzyme in saliva that initiates carbohydrate digestion is lipase.

1.3 _____ The process of mastication results in smaller particles of food that are easier to swallow and easier to break down chemically in the digestive tract.

1.4 _____ Another word that is sometimes used for swallowing is deglutition.

1.5 _____ The esophagus is composed of two layers of smooth muscle that contract during peristalsis to transport food to the stomach. The orientation of the outer layer of muscle fibers is circular, and the inner layer is longitudinal.

1.6 _____ Acetic acid is produced by the stomach; it helps to break down the bolus of food into chyme, a more liquid substance.

1.7 _____ The stomach enzyme pepsin initiates protein digestion of food.

1.8 _____ Most chemical digestion of food occurs in the stomach.

1.9 _____ The total length of the small intestine is about 12 feet.

1.10 _____ Bile and pancreatic enzymes empty into the jejunum portion of the small intestine.

1.11 _____ Some of the enzymes produced by the pancreas to aid in chemical digestion of food include lipase, trypsin, and amylase.

1.12 _____ Finger-like projections of the intestinal wall that increase the surface area of the wall for more efficient food absorption are called rugae.

1.13 _____ Breakdown products of carbohydrate, protein, and fat digestion are absorbed through the capillary walls of blood vessels located within the intestinal villi.

1.14 ______ The functional cell of the liver that detoxifies blood of harmful substances and stores vitamins and glucose is the parietal cell.

1.15 ______ One of the functions of the appendix is to aid in the body's defense against invasion by pathogenic microorganisms that enter the digestive tract.

1.16 ______ The major activity that occurs in the large intestine is the absorption of water and minerals from undigested food, which results in the production of feces.

Exercise 2

Look at the image in the Primary window. See how many of the structures of the digestive system you can identify without looking at the labels. After quizzing yourself, click the Label button and see how many of the structures you identified correctly. After you have identified a structure, listen to the correct pronunciation of the name, and practice saying it if you do not know how to pronounce it correctly.

Exercise 3

After completing your review of the overview text and without looking at the image in the Primary window, list, in order from proximal to distal, the components of the digestive tract, beginning with the mouth and ending with the anus.

ORAL CAVITY AND PHARYNX

Exercise 4

Using View in the Menu bar, select Full Anatomy. Choose the Medial view, and examine the Tongue layer in the Structure List window. As you may remember from our study of the respiratory system, the oral cavity is separated from the nasal cavity by the palate, which is composed of bony and muscular parts. Magnify the structures in the oral cavity and pharynx using the Zoom button, and review the regions before you answer the following questions.

4.1 Which part of the soft palate "hangs" down into the oral cavity and functions to separate the nasopharynx and the oropharynx?

4.2 Name the large muscle that attaches the tongue to the genu of the anterior mandible.

4.3 Name the intrinsic tongue muscle that is located on the superior part of the tongue.

4.4 Which structures lie on the surface of the tongue and are used for tasting?

4.5 Which part of the digestive system continues inferiorly below the laryngopharynx?

TEETH

Exercise 5

Draw a simple sketch of a set of permanent teeth in the upper jaw (16 teeth total). In your drawing, label the different types of teeth, including the incisors, canines, premolars, and molars. These types can be identified in both the Anterior and Lateral views of the Skull layer from the Structure List window.

SALIVARY GLANDS

Exercise 6

Three pairs of salivary glands produce secretions outside the oral cavity, which are added to the digestive tract inside the oral cavity. One pair of glands, the parotid glands, are located on the lateral side of the head, and can be seen when the Lateral view is selected and the Parotid Gland and Duct layer is chosen from the Structure List window. You may want to use the Zoom button to enlarge the area anterior to the ear, where the parotid gland is located.

6.1 The parotid gland overlies which muscle of mastication that inserts on the mandible?

6.2 The secretions of the parotid gland must be transported from the gland to the mouth, where they aid in the process of digestion. The parotid duct, which transports these secretions, crosses the masseter muscle, goes through the buccal fat pad, and enters the mouth through the cheek. Follow the pathway of the parotid duct and name the muscle that it penetrates to enter the oral cavity.

6.3 Which cranial nerve has several branches that travel through the parotid gland to innervate the muscles of facial expression?

Exercise 7

Two other pairs of salivary glands also produce secretions that are added to the digestive tract to aid in digestion of food. These can be seen in the area below the tongue when you select the Deep Salivary Glands layer from the Structure List window in the Lateral view.

7.1 One of these glands has been cut in this dissection. Which one?

7.2 Name the other salivary gland seen in this view.

7.3 Name two nerves that have close relationships to these two salivary glands.

a.

b.

ESOPHAGUS, STOMACH, SMALL INTESTINE, AND LARGE INTESTINE

Exercise 8

The esophagus begins at the inferior border of the laryngopharynx, descends through the thorax, travels through the diaphragm, and opens into the stomach in the cardiac region. The cranial nerve that travels with it furnishes parasympathetic innervation to smooth muscle, cardiac muscle, and the glands of the thorax, abdomen, and pelvis. Its anterior trunk can be seen when you select the Anterior view and choose Stomach from the Structure List window. Use the Zoom button if you need to enlarge the image.

8.1 Name the cranial nerve.

Exercise 9

The stomach is a C-shaped organ that appears to be smooth on its surface. While in the Anterior view and Stomach layer, review the gross anatomy of the stomach, including its curvatures and regions. Now you are going to perform surgery and open the stomach to see its internal anatomy. Open the OR Panel by clicking the OR button in the Primary window. Click and hold the Scalpel button and select Wide for the incision width. Make an incision across the length of the stomach to expose its internal surface.

9.1 Describe the anatomy of the internal stomach.

9.2 Which word is used to describe this type of folding arrangement?

9.3 What is the purpose of this kind of surface? (*Hint: You learned about this when you read the text overview information in Exercise 1.*)

Exercise 10

To view the remainder of the digestive tract, examine the Transverse Colon layer and the Ascending and Descending Colon layer in the Structure List window. Carefully study the gross anatomy of the stomach, small intestine, and large intestine in all of these views.

Listed on the top row in the table below are the regions of the digestive tract you have just reviewed. In the left column are several anatomical structures, or landmarks. Make a check mark in the box that matches the structure with the region where it is found. If you have trouble locating a particular structure, you can use the Find button to help you.

	Stomach	Small Intestine	Large Intestine
10.1 taenia coli			
10.2 pyloric sphincter			
10.3 jejunum			
10.4 lesser curvature			
10.5 mesentery			
10.6 epiploic appendages			
10.7 fundus			
10.8 hepatic flexure			
10.9 ileum			
10.10 vermiform appendix			
10.11 duodenum			
10.12 cecum			
10.13 greater curvature			
10.14 splenic flexure			
10.15 body			

LIVER, GALL BLADDER, AND PANCREAS

Exercise 11

The liver, gall bladder, and pancreas are accessory organs that are not part of the digestive tract, but which produce important substances that are added to the tract in the duodenum. The liver and gall bladder can be examined by selecting the Anterior view and choosing the Liver layer and the Gall Bladder layer in the Structure List window.

11.1 Which muscle tendon lies immediately superior to the liver?

11.2 As you can see, the lobes of the liver are not equal in size. Which is larger?

11.3 Name the ligament that lies between the two lobes.

11.4 Which structure is located in the inferior border of the ligament described in 11.3?

11.5 Which large vein lies superior and posterior to the liver in this view?

11.6 Which vein can be seen entering the liver on its undersurface?

11.7 Name the artery and its two branches that supply the liver, that can also be identified in the same region described in 11.6.

11.8 Draw a simple sketch and label the ductal system that transports bile from the liver and gall bladder to the duodenum. Include the following structures in your drawing: liver, gall bladder, duodenum, right and left hepatic ducts, common hepatic duct, cystic duct, and common bile duct.

Exercise 12

The anatomy of the pancreas and its relationships to other organs and structures can be seen in the Anterior view by selecting the Pancreas layer from the Structure List window.

12.1 Describe the relationship of the head of the pancreas to the duodenum.

12.2 Name the lymphatic organ that lies very close to the tail of the pancreas.

12.3 Which organs lie posterior to the pancreas on both sides?

12.4 Name the two vessels that are located inferior to the pancreas and anterior to the duodenum.

a.

b.

PERITONEUM

Exercise 13

Some of the abdominal organs of the digestive system are contained within a double-walled membrane called the peritoneum. This serous membrane has two layers, the parietal and visceral peritoneum, with the peritoneal cavity between. The parietal peritoneum is the outer layer that lines the abdominal cavity and can be identified by selecting the Anterior view and choosing the Parietal Peritoneum layer from the Structure List window. Click the Isolate button and Highlight the peritoneum to see its location in the abdomen. The inner layer, the visceral peritoneum, closely adheres to the surface of many of the organs. Organs that have peritoneum only on their anterior surfaces are called retroperitoneal. Peritoneal reflections are sheets or folds of peritoneum that anchor organs to the posterior body wall or to other organs.

13.1 One of these peritoneal reflections is the greater omentum. Remove the parietal peritoneum layer by choosing the next layer, the Greater Omentum layer, from the Structure List window. Click the Isolate button to see it more clearly. Describe its attachment to the stomach and its position in the abdomen.

13.2 Another peritoneal reflection is the lesser omentum, which can best be seen in the Lesser Omentum layer. Isolate this structure and describe its attachments.

13.3 The falciform ligament is another reflection that attaches an organ we have previously examined to the anterior body wall and diaphragm. What is that organ?

Exercise 14

Two other examples of peritoneal reflections can be identified by selecting the Medial view and choosing the Mesentery layer from the Structure List window.

14.1 Identify the transverse colon in this view. Name the structure that attaches it to the posterior body wall.

14.2 By what structure are the jejunum and ileum attached to the posterior abdominal wall?

10

URINARY SYSTEM

STUDENT OBJECTIVES

Overview

- Review basic information on the anatomy and physiology of the urinary system.

Location and External Anatomy of the Kidney

- Name two supporting structures that help maintain the normal position of the kidneys in the body.
- Describe the location of the kidneys and their relationships to other structures in the region.

Renal Vessels and Ureters

- Name the arteries that supply blood to the kidneys and the veins that drain them.
- Describe the course of the ureter from the kidney to the bladder.

Internal Anatomy of the Kidney

- Describe the internal anatomy of the kidney.
- Trace the pathway of urine flow from its site of production in the kidney to the urinary bladder.

Bladder and Urethra

- Describe the position of the bladder in both the female and the male.
- Name the three portions of the male urethra.

OVERVIEW

Exercise 1

Select View in the Menu bar and choose Urinary System. Review both the male and the female anatomy by clicking Options in the Menu bar. Read the text overview and complete the following statements about the structures and function of the urinary system.

1.1 The kidneys are located in the lower back, near the ________________ rib.

1.2 The darker, inner part of the kidney, which contains pyramids, is the ________________.

1.3 The functional unit of the kidney is the ______________________.

1.4 A group of ____________________ constitutes the glomerulus.

1.5 The glomerular capsule and glomerulus form the ____________________.

1.6 The filtration barrier that surrounds the capillaries is formed from "foot" processes called ____________________.

1.7 Some of the substances that are small enough to pass through the filtration barrier include ____________________.

1.8 The name of the liquid substance formed during the filtration process is ____________________.

1.9 During the production of urine, several substances useful to the body are reabsorbed into the bloodstream. These include ____________________ .

1.10 ____________________ is the hormone released from the pituitary gland that reduces urine volume by permitting more water to be reabsorbed into the bloodstream by the collecting ducts.

1.11 Another term sometimes used for urination is ____________________.

1.12 The hormone that increases blood volume and blood pressure by increasing sodium reabsorption and, therefore, water reabsorption from the collecting tubules is ____________________.

1.13 The hormone ____________________ is produced by the kidneys and controls erythrocyte production in the bone marrow.

LOCATION AND EXTERNAL ANATOMY OF THE KIDNEY

Exercise 2

The kidneys are supported by connective tissues that help to keep them in their normal position in the abdomen. Using View in the Menu bar, select Full Anatomy. Choose the Anterior view and the Kidney layer from the Structure List window.

2.1 Name the outermost covering that anchors the kidneys to surrounding tissues. (You can also see this layer when you look at the Lateral view, Perirenal Fat and Renal Fascia layer.)

2.2 Which "yellow" connective tissue mass lies beneath this covering and helps protect the kidneys?

2.3 Which kidney is superior in position in the abdominal cavity?

2.4 Name the large organ in the upper right quadrant that occupies space in the abdomen and prevents the kidneys from lying at the same level on the posterior body wall.

2.5 Identify and name the endocrine gland that "sits" on the superior part of the kidney.

2.6 Which muscle lies posterior and inferior to each kidney?

RENAL VESSELS AND URETERS

Exercise 3

The blood vessels of the kidney and the ureters that transport urine to the bladder can be reviewed by scrolling down the layers in the Structure List window from the Kidney layer to the Inferior Vena Cava layer, and then to the Abdominal Aorta and Branches layer.

3.1 Name the veins that drain blood from the kidneys into the inferior vena cava.

3.2 Are they the same length? Which is longer?

3.3 The region of the kidney where the renal vein exits and the renal artery enters is the ________________. (*Hint: You should review the Kidney layer to answer this question.*)

3.4 Which of the following structures is most anterior in position as it leaves or enters the kidney? (circle the correct answer)

renal artery, ureter, renal vein

3.5 Describe the relationship of the ureters to the iliac arteries and veins.

INTERNAL ANATOMY OF THE KIDNEY

Exercise 4

While in the Anterior view, use the Zoom button to study the internal anatomy of the kidney. Select the Kidney—Longitudinal Section layer from the Structure List window. Draw a rough sketch of the kidney and label the following structures in your drawing.

renal cortex	renal papilla	minor calyx
renal column	renal sinus	major calyx
renal pyramid	renal pelvis	ureter

Exercise 5

Circle the correct answers to the following questions. Your drawing should help you with the answers to these questions.

5.1 The correct pathway of urine flow from its site of production to the urinary bladder is:

a. renal medulla → renal sinus → minor calyx → major calyx → renal cortex → renal pelvis → ureter → urinary bladder

b. renal cortex → renal pyramid → renal papilla → minor calyx → major calyx → renal pelvis → ureter → urinary bladder

c. renal cortex → renal pyramid → renal pelvis → minor calyx → major calyx → renal papilla → ureter → urinary bladder

d. renal column → renal pyramid → renal papilla → minor calyx → major calyx → renal pelvis → ureter → urinary bladder

5.2 Approximately how many minor calyces drain into each major calyx?

a. 4–6
b. 5–8
c. 2–3
d. 1–2

5.3 Portions of the renal cortex that extend between the renal pyramids in the medulla are the ____________________.

a. renal columns
b. renal sinuses
c. renal papillae
d. minor calyces

5.4 The apical part of the renal pyramid is the ____________________.

a. cortex
b. major calyx
c. renal column
d. renal papilla

5.5 The ________________ collects urine from the major calyces and drains into the ureter.

a. renal pelvis
b. renal column
c. renal sinus
d. renal papilla

BLADDER AND URETHRA

Exercise 6

The bladder receives urine from the ureters and stores it until it is released from the body through the urethra. The normal positions of the bladder and urethra in the female can be seen by choosing the Medial view and the Urinary Bladder layer from the Structure List window. In the male, select the same view and the Urinary Bladder and Prostate Gland layer.

6.1 In both the female and the male, the urinary bladder lies posterior to which structure?

6.2 List the three portions of the male urethra based on their anatomical location.

a.

b.

c.

6.3 Which muscle surrounds the membranous portion of the male urethra?

11

REPRODUCTIVE SYSTEM

STUDENT OBJECTIVES

Overview

- Review basic information on the anatomy and physiology of the reproductive system.
- Discuss the site of production of the major hormones of the reproductive system.

Male Reproductive System

- Describe the layers of tissue, from superficial to deep, that form the scrotum.
- Name several structures that are contained within the spermatic cord.
- Describe the formation of the inguinal ligament and the anatomy of the superficial inguinal ring.
- Name all of the structures, in order, that transport sperm from their site of production in the seminiferous tubules to their release from the penis.
- Describe the anatomical relationships of the seminal vesicle and the prostate gland to the bladder.
- Name the two supporting ligaments of the penis.
- Describe the relationships of the deep vein, arteries, and nerves on the dorsum of the penis.
- Name two of the three erectile bodies of the penis that can be seen in Medial view, and describe the course of the urethra through the penis.

Female Reproductive System

- Describe the location of the broad ligament and the ovarian ligament.
- Describe the arterial supply and venous drainage of the uterus and ovaries.
- Discuss the location and structure of the ovaries, uterine tubes, uterus, vagina, and vulva.
- Discuss the location and structure of the mammary glands.
- Describe the arteries, nerves, and lymphatic vessels that have close relationships to the mammary glands.

OVERVIEW

Exercise 1

Click on View in the Menu bar, select the Reproductive System, read the text overview, and look at the image in the Primary window. You can look at both the male and the female images by clicking the Options button and selecting the image you want. After reviewing the images and the text, circle the correct answer to the following questions.

1.1 The site of production of sperm in the male is the ________________.

a. Leydig cells
b. seminiferous tubules
c. sertoli cells
d. none of the above

1.2 Which portion of the mature sperm contains mitochondria, which provides energy for sperm motility?

a. flagellum
b. acrosome
c. tail
d. middle piece

1.3 How many sperm are released during an ejaculation?

a. 50–100 million
b. 100–200 million
c. 300–500 million
d. 500–800 million

1.4 In which part of the male reproductive tract does storage and maturation of sperm occur?

a. seminiferous tubule
b. prostate gland
c. seminal vesicle
d. epididymis

1.5 Stimulation of which type of nerve fibers in the vas deferens produces peristaltic muscle contractions that propel sperm into the ejaculatory duct?

a. somatic motor nerves
b. parasympathetic nerves
c. sympathetic nerves
d. none of the above

1.6 Which accessory gland produces secretions that provide energy for the sperm and also a neutralizing chemical that reduces vaginal acidity?

a. penis
b. bulbourethral gland
c. prostate gland
d. seminal vesicle

1.7 When circumcision is performed surgically, which part of the penis is removed?

a. erectile tissue
b. glans
c. prepuce
d. shaft

1.8 Which part of the female reproductive tract contains developing follicles?

a. vagina
b. uterine tube
c. ovary
d. uterus

1.9 In addition to the smooth muscle in the walls of the uterine tubes, which other structures in the walls help to move the ova through the tubes?

a. microvilli
b. cilia
c. centrioles
d. microfilaments

1.10 The normal position of the uterus in the pelvic cavity is ________________.

a. prolapsed
b. inclined backward
c. anteverted
d. none of the above

1.11 Which layer of the uterus contains the stratum functionalis?

a. stratum basalis
b. myometrium
c. serous layer
d. endometrium

1.12 Which of the following is not part of the vulva in the female?

a. labia majora
b. labia minora
c. cervix
d. mons pubis

1.13 What is the erectile tissue in the female that is the homologue of the penis in the male?

a. clitoris

b. labia majora

c. labia minora

d. mons pubis

1.14 Approximately how many lobes of glandular tissue are contained in each breast?

a. 5–10

b. 15–20

c. 30–40

d. 50–60

1.15 Which structures help to support the breasts by connecting them to the skin?

a. suspensory ligaments

b. lactiferous ducts

c. lactiferous sinuses

d. none of the above

Exercise 2

After reading the text overview, you should be able to trace the pathway that sperm travel through the male reproductive tract. List all of the structures, in order, that transport sperm from their site of production in the seminiferous tubules to their release from the penis.

Exercise 3

Match the hormones listed below with their site of production. You should complete this exercise after you have read the text overview.

_____	3.1 estrogen	a. anterior pituitary
_____	3.2 testosterone	b. posterior pituitary
_____	3.3 FSH	c. ovary
_____	3.4 Gn-RH	d. testis
_____	3.5 progesterone	e. hypothalamus
_____	3.6 LH	f. breast

MALE REPRODUCTIVE SYSTEM

Scrotum and Spermatic Cord

Exercise 4

Using View in the Menu bar, select Full Anatomy. Click Options, and choose the Medial view and Male. Select the Ductus Deferens layer from the Structure List window. The scrotum is the sac-like structure of the male reproductive system that contains the testes, or testicles. Several layers of tissue, including two muscles, contribute to the formation of this sac. Arrange the following layers of tissue as they are located from superficial to deep within the scrotum. You will need to use the Zoom button and enlarge the image.

skin and dartos muscle	cremaster muscle
internal spermatic fascia	skin
external spermatic fascia	superficial scrotal fascia
tunica vaginalis	

4.1 ______________________ 4.5 ______________________

4.2 ______________________ 4.6 ______________________

4.3 ______________________ 4.7 ______________________

4.4 ______________________

Exercise 5

The same layers of tissue that are part of the scrotal sac are also components of the spermatic cord. Select the Anterior view, and review the Skin layer and Cutaneous Nerve layer from the Structure List window.

5.1 A branch of which cutaneous nerve travels within the scrotal sac to innervate the skin of the scrotum?

Now select the External Abdominal Oblique Muscle layer and answer the following questions about the inguinal canal region.

5.2 The inferior border of the aponeurosis of which muscle forms the inguinal ligament?

5.3 Where does the spermatic cord enter the inguinal canal?

5.4 Describe the relationships of the medial crus, lateral crus, and intercrural fibers to the superficial inguinal ring.

Scroll down the Structure List window and examine the Internal Abdominal Oblique Muscle layer.

5.5 From which abdominal muscle does the cremaster muscle arise?

Exercise 6

The spermatic cord contains several structures in addition to the layers examined in Exercises 4 and 5. While still in the Anterior view, choose the Gonadal Veins layer and the Testes layer from the Structure List window.

6.1 Name three structures that can be identified in the spermatic cord.

a.

b.

c.

6.2 Follow the ascending root of the pampiniform plexus from the testes, and state which vessel this plexus of veins drains into.

Ducts and Accessory Glands

Exercise 7

The ducts that transport sperm from the testes, and some of the accessory glands that add secretions to the sperm to form semen, can be seen in Medial view by choosing the Urinary Bladder and Prostate Gland layer from the Structure List window.

7.1 Identify the epididymis, which is located on the superior part of the testis. The ductus deferens transports sperm from the epididymis and travels with other structures in the spermatic cord through the inguinal canal. Follow the course of the ductus deferens after it enters the pelvic cavity. Now identify the expanded part of the ductus deferens that is located between the bladder and the rectum. What is it?

7.2 Describe the anatomical relationships of the following accessory glands to the bladder.

a. seminal vesicle

b. prostate gland

7.3 Name the duct located in the prostate gland that is formed by the junction of the ductus deferens and the duct from the seminal vesicle.

Penis

Exercise 8

The external anatomy of the penis can be reviewed when you select the Anterior view and examine several different layers, from superficial to deep. Begin your review by scrolling down layer by layer through the Structure List window, from the Skin layer to the Cutaneous Nerves layer. Then examine the Deep Fascia of the Penis and the Deep Dorsal Vein of the Penis layers, and answer the following questions.

8.1 What is the name of the enlarged tip of the penis?

8.2 Two ligaments help to support the weight of the penis, and can be identified in the superficial and deep views.

a. Which ligament is more superficial and arises from the inferior part of the linea alba? (*See the Cutaneous Nerves layer.*)

b. Which ligament is deeper and attaches to the deep fascia of the penis? (*See Deep Fascia of the Penis layer.*)

8.3 Describe the relationships of the deep dorsal vein, dorsal arteries, and dorsal nerves on the dorsum of the penis. (*See the Deep Dorsal Vein of the Penis layer.*)

Exercise 9

The internal anatomy of the penis can be seen using the Medial view and selecting the Urinary Bladder and Prostate Gland layer from the Structure List window.

9.1 Two of the three erectile bodies of tissue that form the penis can be seen in this view. Name them.

a.

b.

9.2 Through which of the erectile bodies listed in 9.1 does the urethra travel?

9.3 Name the muscle that surrounds the bulb of the penis.

FEMALE REPRODUCTIVE SYSTEM

Ovaries, Uterine Tubes, Uterus, Vagina, and Vulva

Exercise 10

The structures of the female reproductive system can be reviewed when you choose Full Anatomy from View on the Menu bar and select Anterior view and Female from Options. Examine the Peritoneum layer from the Structure List window.

10.1 Connective tissue structures help to support and anchor the female reproductive organs in their normal position in the pelvic cavity. What is the extensive "sheet-like" peritoneal structure that overlies the uterus and other parts of the reproductive system?

Exercise 11

When you change to the Uterus layer, the broad ligament has been removed and you can see the female reproductive organs more clearly.

11.1 Identify and name the ligament that anchors each ovary to the uterus.

11.2 Draw an anterior view of the uterus, uterine tubes, and ovaries in their normal anatomical position in the pelvic cavity. Include in your drawing the following labels: the fundus, the body and cervix of the uterus, the ovary, and the isthmus, ampulla, infundibulum, and fimbriae of the uterine tubes.

Exercise 12

To examine the vessels that supply and drain the structures described in Exercise 11.2, select the Ovarian Veins layer from the Structure List window.

12.1 List the arteries and veins of the following structures:

a. ovary

b. uterus

Exercise 13

The organs of the female reproductive system can also be seen in the Medial view by choosing the Uterus layer from the Structure List window. You may want to use the Zoom button to enlarge the image.

13.1 Describe the relationship of the fornix of the vagina to the cervix.

13.2 Which muscle surrounds the urethra and the vagina and is part of the urogenital diaphragm?

13.3 Which structure of the vulva (external genitalia) encloses the openings of the urethra and the vagina?

13.4 Name the body of erectile tissue located anterior to the urethra near the pubic symphysis.

Mammary Glands

Exercise 14

The anatomy of the mammary glands (breasts) can be examined in several layers by selecting the Anterior view and scrolling down the Structure List window from the Skin layer to the Breast layer. Answer T if the statement is true and F if the statement is false.

14. 1 _____ The pigmented area that surrounds the nipple is the areola.

14.2 _____ The breasts contain adipose tissue and overlie the pectoralis major muscle fascia.

14.3 _____ Each mammary gland is composed of several lobes separated by suspensory ligaments.

14.4 _____ The mammary ducts open to the outside of the breast at the nipple.

14.5 _____ Branches of the superficial epigastric arteries are located very close to the breast, and probably supply them with oxygenated blood.

14.6 _____ Lymphatic vessels travel from each breast toward the midline and the axillary region.

14.7 _____ Nerves that innervate the breast are branches of the brachial plexus.

ANSWER KEY

CHAPTER 1

Exercise 1

1.1 c	1.2 d	1.3 b	1.4 d	1.5 c
1.6 c	1.7 d	1.8 a	1.9 c	1.10 b

Exercise 2

self-quizzing and pronunciations

Exercise 3

3.1 nasal (F), ethmoid (C), maxilla (F), frontal (C), sphenoid (C), zygomatic (F), vomer (F), lacrimal (F), temporal (C), mandible (F)

3.2 frontal, sphenoid, lacrimal, maxilla, zygomatic

3.3 ethmoid, vomer

3.4 1. b 2. a 3. e 4. c 5. d 6. b 7. a 8. d 9. b

Exercise 4

4.1 F strike out ligamentum nuchae, replace with anterior longitudinal ligament

4.2 F strike out smaller, replace with larger

4.3 F strike out axis, replace with atlas

4.4 T

4.5 F strike out median, replace with lateral

4.6 F strike out interspinous, replace with supraspinous

4.7 T

4.8 F strike out annulus fibrosus, replace with nucleus pulposus

Exercise 5

5.1 manubrium, body

5.2 xiphoid

5.3 a. second rib attaches directly at the sternal angle (manubriosternal junction) by its costal cartilage; true rib

b. fifth rib attaches directly to the body by its costal cartilage; true rib

c. tenth rib attaches indirectly to the body by the costal cartilages immediately above; false rib

Exercise 6

6.1 head   6.2 yes   6.3 tubercle

Exercise 7

7.1 c	7.2 h	7.3 f	7.4 e	7.5 b	7.6 c
7.7 d	7.8 b	7.9 a	7.10 c	7.11 b	7.12 e
7.13 f	7.14 c	7.15 a			

7.16 drawing

Exercise 8

8.1 pubic symphysis   8.2 sacrum   8.3 acetabulum

8.4
1. ischium
2. pubis
3. ilium, pubis
4. ilium
5. ischium
6. ilium
7. ischium, pubis
8. ilium
9. ilium

Exercise 9

9.1 drawing

9.2
1. 7
2. calcaneus, talus, navicular, cuboid, medial cuneiform, intermediate cuneiform, lateral cuneiform
3. intermediate cuneiform
4. fourth and fifth
5. 14
6. great toe

Exercise 10

10.1
1. 50%
2. drawing

10.2 drawing

CHAPTER 2

Exercise 1

1.1 c	1.2 d	1.3 b	1.4 d	1.5 c
1.6 b	1.7 b	1.8 d	1.9 d	1.10 c

Exercise 2

2.1 suture: fibrous, synarthrosis

2.2 gomphosis: fibrous, synarthrosis

2.3 syndesmosis: fibrous, amphiarthrosis

2.4 symphysis: cartilaginous, amphiarthrosis

2.5 ball-and-socket: synovial, diarthrosis
2.6 hinge: synovial, diarthrosis

Exercise 3

3.1 suture
3.2 a. coronal suture b. squamosal suture c. lambdoidal suture d. sagittal suture
3.3 gomphosis
3.4 synarthrosis

Exercise 4

4.1 interosseous membrane
4.2 amphiarthrosis

Exercise 5

5.1 synarthrosis

Exercise 6

6.1 a. intervertebral discs b. nucleus pulposus and annulus fibrosus
6.2 pubic symphysis
6.3 amphiarthrosis

Exercise 7

7.1 coracohumeral and glenohumeral ligaments
7.2 supraspinatus and subscapularis muscles
7.3 deltoid and supraspinatus muscles
7.4 long head of biceps brachii muscle

Exercise 8

8.1 radial and ulnar collateral ligaments
8.2 head of radius

Exercise 9

9.1 iliofemoral: Y-shaped ligament attaching between the ilium and the femur
pubofemoral: ligament attaching between the pubis and the femur
ischiofemoral: ligament attaching between the ischium and the femur
9.2 a. articular cartilage b. acetabular labrum

Exercise 10

10.1 rectus femoris, vastus lateralis, vastus medialis
10.2 vastus intermedius
10.3 infrapatellar fat pad
10.4 gracilis, sartorius, semitendinosus

Exercise 11

11.1 tibial (medial) collateral ligament attaches between the medial epicondyle of femur and the medial side of tibia below the medial condyle; fibular (lateral) collateral ligament attaches between the lateral epicondyle of femur and the head of fibula
11.2 anterior cruciate ligament
11.3 genicular arteries and veins

Exercise 12

12.1 drawing
12.2 medial meniscus is fused to the tibial collateral ligament

CHAPTER 3

Exercise 1

1.1 F strike out 206, replace with 600; strike out 10%, replace with 40%
1.2 T
1.3 F strike out heart, replace with glands
1.4 F strike out skeletal, replace with smooth
1.5 F strike out epimysium, replace with perimysium
1.6 T
1.7 F strike out actin, replace with myosin *or* strike out thick, replace with thin
1.8 T
1.9 F strike out ligaments, replace with tendons
1.10 T
1.11 F strike out slow, replace with fast

Exercise 2

Skeletal muscle fibers and nuclei are arranged in parallel rows; cardiac muscle fibers and nuclei are not arranged in parallel rows, and the fibers appear to have a branching pattern with more connective tissue between the fibers.

Exercise 3

self-quizzing and pronunciation

Exercise 4

4.1 j	4.2 f	4.3 a	4.4 g	4.5 b
4.6 h	4.7 c	4.8 i	4.9 d	4.10 e

Exercise 5

5.1 frontalis and occipitalis
5.2 galea aponeurotica
5.3 buccinator
5.4 masseter

Exercise 6

6.1 omohyoid 6.2 sternocleidomastoid 6.3 sternohyoid 6.4 scalenes
6.5 posterior belly 6.6 sternothyroid 6.7 stylohyoid

Exercise 7

7.1 trapezius 7.2 supraspinatus 7.3 teres major 7.4 latissimus dorsi
7.5 splenius capitis 7.6 rhomboideus major 7.7 levator scapulae
7.8 semispinalis capitis 7.9 deltoid 7.10 teres minor
7.11 rhomboideus minor 7.12 infraspinatus

Exercise 8

8.1 scapula 8.2 subscapularis
8.3 serratus posterior superior and serratus posterior inferior
8.4 spinalis group: most medial in position
longissimus group: intermediate in position
iliocostalis group: most lateral in position
8.5 maintains upright posture of the vertebral column

Exercise 9

9.1
a. originates on transverse processes and inserts on spinous processes 4-6 levels above
b. originates on transverse processes and inserts on spinous processes 2-4 levels above
c. originates on transverse processes and inserts on spinous processes 1-2 levels above
d. originates on transverse processes and inserts on ribs 1-2 levels below
e. attaches between adjacent spinous processes
f. attaches between adjacent transverse processes

9.2 rectus capitis posterior minor, rectus capitis posterior major, obliquus capitis inferior and obliquus capitis superior

Exercise 10

10.1 flexion, adduction, and medial rotation of the humerus at the shoulder joint
10.2 stabilizes and depresses the clavicle
10.3 draws scapula forward and downward
10.4 protracts and holds scapula against the chest wall

Exercise 11

11.1 fibers go downward and forward from the inferior border of the rib above to the superior border of the rib below
11.2 laterally
11.3 external intercostal membrane

Exercise 12

12.1 internal intercostal muscles

12.2 fibers are at right angles to the fibers of the external intercostal muscles and go from the superior border of the rib below to the inferior border of the rib above
12.3 yes
12.4 transversus thoracis and innermost intercostal muscles
12.5 innermost intercostal membrane
12.6 second and third layers (internal intercostal and transversus thoracis/innermost intercostal)
12.7 internal thoracic artery and the vena comitans of the internal thoracic artery

Exercise 13

13.1 external oblique   13.2 aponeurosis   13.3 linea alba
13.4 internal oblique   13.5 transversus abdominis   13.6 transversely
13.7 pubic bone   13.8 sternum and ribs   13.9 tendinous intersections
13.10 pyramidalis   13.11 internal oblique   13.12 transversus abdominis

Exercise 14

14.1 short head   14.2 coracobrachialis
14.3 crosses over the flexor muscles of the forearm below the elbow
14.4 brachialis   14.5 styloid process of the radius   14.6 flex

Exercise 15

15.1 long head   15.2 anconeus

Exercise 16

16.1 c   16.2 e   16.3 f   16.4 h
16.5 b   16.6 g   16.7 a   16.8 d

Exercise 17

17.1 extensor retinaculum
17.2 extensor carpi radialis longus, extensor carpi ulnaris, extensor digitorum
17.3 extensor carpi ulnaris
17.4 extensor expansions of the digits
17.5 abductor pollicis longus, extensor pollicis brevis, extensor pollicis longus
17.6 extensor indicis
17.7 humerus and ulna

Exercise 18

drawing

Exercise 19

19.1 iliacus   19.2 tensor fasciae latae   19.3 sartorius   19.4 rectus femoris
19.5 adductor magnus   19.6 vastus intermedius   19.7 adductor longus
19.8 vastus lateralis

Exercise 20

20.1 gluteus maximus: origin—ilium, sacrum, coccyx; insertion—femur and iliotibial tract; action—extension of thigh

20.2 gluteus medius: origin—ilium; insertion—femur; action—abducts and medially rotates thigh

20.3 gluteus minimus: origin—ilium; insertion—femur; action—abducts and medially rotates thigh

Exercise 21

21.1 piriformis and superior gemellus

21.2 piriformis, superior gemellus, obturator internus, inferior gemellus, quadratus femoris, obturator externus

Exercise 22

22.1 biceps femoris 22.2 ischial tuberosity 22.3 semitendinosus

22.4 semimembranosus and biceps femoris

Exercise 23

23.1 a. superior extensor retinaculum crosses tendons transversely above the ankle joint
b. inferior extensor retinaculum is Y-shaped and crosses tendons at the ankle joint and on the dorsum of the foot

23.2 tibialis anterior 23.3 extensor expansions of the digits 23.4 peroneus tertius

23.5 extensor hallucis longus

Exercise 24

24.1 peroneus longus and peroneus brevis 24.2 fifth metatarsal

24.3 superior peroneal retinaculum, inferior peroneal retinaculum

Exercise 25

25.1 F 25.2 T 25.3 T 25.4 T 25.5 T 25.6 F

Exercise 26

drawing

CHAPTER 4

Exercise 1

1.1 brain and spinal cord 1.2 perikaryon 1.3 neurotransmitters

1.4 Schwann cells 1.5 corpus callosum 1.6 cerebrum 1.7 sulci

1.8 thalamus, hypothalamus, epithalamus 1.9 gyri 1.10 vermis

1.11 medulla oblongata 1.12 reticular formation 1.13 vertebral canal

1.14 pia mater 1.15 somatic, autonomic 1.16 gray matter 1.17 dorsal

1.18 sympathetic 1.19 parasympathetic

1.20 receptor, sensory neuron, integration center (CNS), motor neuron, effector

Exercise 2

2.1 free nerve endings 2.2 cones 2.3 organ of Corti 2.4 Pacinian corpuscle
2.5 vestibular apparatus 2.6 rods 2.7 proprioceptors

Exercise 3

3.1 Cerebrum: controls language, conscious thought, hearing, somatosensory function, memory, personality development and vision.

3.2 Hypothalamus: regulates some endocrine glands, visceral activities, food and water intake, sleep and wake patterns, sex drive, and emotional states.

3.3 Thalamus: is the relay and preprocessing station for many nerve impulses.

3.4 Epithalamus: is involved in cerebrospinal fluid production.

3.5 Cerebellum: controls balance, posture, and coordination.

3.6 Midbrain: is the center for visual and auditory reflexes.

3.7 Pons: bridges the cerebellum and higher brain centers with the spinal cord.

3.8 Medulla oblongata: contains control centers for swallowing, breathing, digestion, and heartbeat.

3.9 Spinal cord: carries messages between the central nervous system and the rest of the body, and mediates spinal reflexes.

Exercise 4

self-quizzing and pronunciation

Exercise 5

5.1 precentral gyrus, postcentral gyrus 5.2 lateral sulcus
5.3. frontal lobe, occipital lobe 5.4 temporal lobe, parietal lobe

Exercise 6

6.1 midsagittal 6.2 crista galli 6.3 superior sagittal sinus, inferior sagittal sinus, straight sinus 6.4 confluence of sinuses

Exercise 7

7.1 d 7.2 f 7.3 a 7.4 b 7.5 d 7.6 c
7.7 a 7.8 d

Exercise 8

8.1 midbrain 8.2 pons, medulla oblongata, cerebellum 8.3 cerebrum 8.4 thalamus and hypothalamus

Exercise 9

drawing

Exercise 10

temporal, zygomatic, buccal, mandibular and cervical branches

Exercise 11

11.1 trochlear nerve—CN IV; trigeminal nerve—CN V; abducent (abducens) nerve—CN VI; glossopharyngeal nerve—CN IX; vagus nerve—CN X
11.2 glossopharyngeal nerve—CN IX
11.3 vagus nerve—CN X
11.4 ophthalmic, maxillary, and mandibular divisions

Exercise 12

12.1 foramen magnum 12.2 dura mater 12.3 arachnoid mater
12.4 cervical region, lumbar region 12.5 termination of spinal cord at about the L1-L2 levels
12.6 filum terminale

Exercise 13

13.1 dorsal, ventral 13.2 cauda equina 13.3 dorsal root ganglion
13.4 dorsal, ventral 13.5 ventral 13.6 dorsal
13.7 communicating 13.8 31 13.9 8, 12, 5, 5, 1 13.10 above
13.11 below 13.12 plexuses

Exercise 14

transverse cervical and supraclavicular nerves

Exercise 15

15.1 lesser occipital and great (greater) auricular nerves
15.2 C2 and C3 nerves
15.3 C3, C4, and C5 nerves

Exercise 16

drawing

Exercise 17

17.1 a. ventral rami of C5 and C6 b. ventral ramus of C7 c. ventral rami of C8 and T1
17.2 axillary artery 17.3 lateral cord 17.4 coracobrachialis muscle
17.5 biceps brachii and brachialis muscles 17.6 lateral and medial cords
17.7 none 17.8 brachial artery and vein 17.9 medial cord
17.10 flexor carpi ulnaris and flexor digitorum profundus muscles 17.11 ulnar nerve
17.12 axillary and radial nerves
17.13 lateral pectoral nerve, medial pectoral nerve, suprascapular nerve, long thoracic nerve, and medial cutaneous nerve of the forearm (medial antebrachial cutaneous nerve)

Exercise 18

drawing

Exercise 19

superior gluteal nerve, inferior gluteal nerve, sciatic nerve, posterior cutaneous nerve of the thigh (posterior femoral cutaneous nerve), and nerve to the superior gemellus and obturator internus muscles

Exercise 20

20.1 biceps femoris and semitendinosus muscles
20.2 common peroneal and tibial nerves
20.3 superficial and deep peroneal nerves
20.4 a. tibial nerve
b. superficial peroneal nerve
c. deep peroneal nerve

Exercises 21 and 22

drawings

CHAPTER 5

Exercise 1

1.1 b 1.2 d 1.3 c 1.4 c 1.5 b
1.6 d 1.7 d 1.8 a 1.9 d 1.10 b

Exercise 2

2.1 PTH 2.2 increases blood calcium 2.3 ACTH 2.4 anterior pituitary
2.5 stimulates thyroid gland to produce thyroid hormones 2.6 estrogen 2.7 thyroid
2.8 affects body growth, metabolic rates and the development of bones and skeletal muscles
2.9 aldosterone 2.10 testosterone 2.11 anterior pituitary
2.12 stimulates protein synthesis and cell division in cartilage and bone tissue
2.13 adrenal medulla 2.14 anterior pituitary 2.15 prolactin 2.16 thyroid
2.17 calcitonin

Exercise 3

identification of glands

Exercise 4

4.1 hypothalamus 4.2 infundibulum
4.3 adenohypophysis (anterior pituitary) and neurohypophysis (posterior pituitary)
4.4 sella turcica 4.5 sphenoid sinus

Exercise 5

5.1 at the level of tracheal rings 2, 3, and 4

5.2 superior thyroid artery from the external carotid artery and inferior thyroid artery from the thyrocervical trunk of the subclavian artery

Exercise 6

6.1 the superior thyroid vein empties into the internal jugular vein; the middle thyroid vein empties into the internal jugular vein; and the inferior thyroid vein empties into the left brachiocephalic vein

Exercise 7

7.1 on the superior pole of each kidney

7.2 the superior, middle, and inferior suprarenal (adrenal) arteries

7.3 the left suprarenal (adrenal) vein drains into the left renal vein; the right suprarenal vein drains into the inferior vena cava

Exercise 8

8.1 the roof of the third ventricle within the diencephalon

CHAPTER 6

Exercise 1

1.1 b	1.2 c	1.3 d	1.4 b	1.5 d
1.6 a	1.7 d	1.8 a	1.9 b	1.10 d

Exercise 2

self-quizzing and pronunciations

Exercise 3

3.1 diaphragm

3.2 a. phrenic nerve b. pericardiacophrenic artery c. pericardiacophrenic vein

Exercise 4

4.1 epicardium of the heart 4.2 pericardial fluid 4.3 fat

Exercise 5

5.1 right atrium 5.2 right ventricle 5.3 left ventricle
5.4 auricle 5.5 superior vena cava 5.6 pulmonary trunk
5.7 and 5.8 left and right pulmonary 5.9 left lung 5.10 posterior
5.11 ascending aorta 5.12 right coronary 5.13 right marginal
5.14 right marginal 5.15 anterior cardiac 5.16 anterior interventricular artery
5.17 great cardiac

Exercise 6

6.1 the coronary sulcus contains the right coronary artery

6.2 the anterior interventricular sulcus contains the anterior interventricular artery and the great cardiac vein

Exercise 7

7.1 fossa ovalis 7.2 left ventricle

7.3 it pumps blood through the systemic circuit

Exercise 8

8.1 the tricuspid valve is located between the right atrium and the right ventricle; the pulmonary (semilunar) valve is located between the pulmonary trunk and the right ventricle

8.2 no 8.3 chordae tendineae

8.4 function is to anchor the valve to the papillary muscle and to keep the valve closed during ventricular contraction

Exercise 9

9.1 two right pulmonary veins, two left pulmonary veins, inferior vena cava

9.2 lungs

9.3 heart, pulmonary trunk, pulmonary arteries, lungs, pulmonary veins, heart

Exercise 10

drawing

Exercise 11

11.1 brachiocephalic trunk, left common carotid artery, left subclavian artery

11.2 right common carotid artery and right subclavian artery

11.3 left brachiocephalic vein 11.4 left vagus nerve

11.5 between the right brachiocephalic vein and the brachiocephalic trunk

Exercise 12

12.1 external and internal carotid arteries

12.2 near the superior border of the larynx

12.3 superficial temporal, maxillary, occipital, facial, lingual, and superior thyroid arteries

Exercise 13

13.1 corpus callosum 13.2 subclavian artery 13.3 basilar artery

13.4 labyrinthine, pontine, superior cerebellar arteries

13.5 cerebral

Exercise 14

14.1 thyrocervical trunk, internal thoracic artery, costocervical trunk, dorsal scapular artery

14.2 as it crosses over the first rib to enter the axilla
14.3 lateral thoracic artery
14.4 anterior and posterior humeral circumflex arteries
14.5 brachial 14.6 deep brachial artery
14.7 superior and inferior ulnar collateral arteries
14.8 brachialis muscle tendon 14.9 radial and ulnar
14.10 ulnar

Exercise 15

15.1 F 15.2 T 15.3 F 15.4 T 15.5 T
15.6 F 15.7 T 15.8 F 15.9 T 15.10 F

Exercise 16

16.1 deep femoral artery 16.2 lateral femoral circumflex artery

Exercise 17

17.1 popliteal artery

Exercise 18

18.1 tibial nerve and popliteal vein
18.2 medial superior and inferior genicular arteries; lateral superior and inferior genicular arteries
18.3 posterior and anterior tibial arteries

Exercise 19

19.1 flexus hallucis longus, tibialis posterior and flexor digitorum longus muscles
19.2 medial plantar and lateral plantar arteries
19.3 distribute arterial blood to the toes

Exercise 20

20.1 fibula 20.2 laterally

Exercise 21

21.1 anterior tibial recurrent artery 21.2 deep peroneal nerve
21.3 dorsalis pedis artery 21.4 medial and lateral tarsal arteries
21.5 five 21.6 two

Exercise 22

22.1 sternocleidomastoid muscle 22.2 subclavian vein
22.3 facial, lingual, and superior thyroid veins
22.4 brachiocephalic vein 22.5 vagus nerve (CN X)

Exercise 23

23.1 basilic and cephalic veins   23.2 median cubital vein

Exercise 24

24.1 radial and ulnar veins, brachial vein, axillary vein, subclavian vein, brachiocephalic vein

24.2 two

Exercise 25

25.1 azygos   25.2 hemiazygos, accessory hemiazygos

25.3 posterior intercostal   25.4 ascending lumbar   25.5 lumbar

Exercise 26

26.1 common iliac veins   26.2 hepatic veins, renal veins

26.3 left

26.4 no; right testicular vein drains directly into the inferior vena cava; left testicular vein drains into the left renal vein

26.5 suprarenal (adrenal) vein   26.6 external iliac vein

Exercise 27

27.1 superior mesenteric vein, splenic vein, inferior mesenteric vein

Exercise 28

28.1 digital veins   28.2 small saphenous vein   28.3 great saphenous vein

Exercise 29

29.1 medial and lateral plantar veins, posterior tibial vein, popliteal vein, femoral vein, external iliac vein

29.2 peroneal vein   29.3 anterior tibial vein

CHAPTER 7

Exercise 1

1.1 T   1.2 T   1.3 F   1.4 T   1.5 F

1.6 T   1.7 T   1.8 T   1.9 T   1.10 T

Exercise 2

2.1
a. crosses anterior to the vertebral column
b. inferiorly it is anterior to the azygos vein, but as it ascends it lies on the left side of the azygos vein
c. lies to the right of the hemiazygos vein

d. lies to the right of the accessory hemiazygos vein
e. crosses over the left sympathetic trunk near the first rib

2.2 left half of the head and neck, left upper limb, left side of the thorax, abdomen, pelvis, lower limbs

Exercise 3

drawing

Exercise 4

4.1 stomach   4.2 diaphragm   4.3 pancreas   4.4 splenic artery
4.5 splenic vein   4.6 left kidney

Exercise 5

5.1 manubrium   5.2 trachea   5.3 left brachiocephalic
5.4 and 5.5 brachiocephalic trunk and left common carotid
5.6 pericardial sac

Exercise 6

6.1 palatoglossus muscle and muscle of the soft palate   6.2 base of tongue

CHAPTER 8

Exercise 1

1.1 d   1.2 b   1.3 c   1.4 a   1.5 b
1.6 a   1.7 c   1.8 b   1.9 a   1.10 d

Exercise 2

nasal cavity → pharynx → larynx → trachea → bronchus → bronchioles → clusters of alveoli (air sacs) → alveoli

Exercise 3

drawing

Exercise 4

middle meatus

Exercise 5

frontal sinus, sphenoidal sinus, ethmoid air cells

Exercise 6

6.1 vestibule   6.2 cribriform plate   6.3 and 6.4 maxilla and palatine

6.5 oral 6.6, 6.7, and 6.8 superior, middle, and inferior
6.9 nasopharynx 6.10 oropharynx 6.11 laryngopharynx
6.12 auditory tube 6.13 pharyngeal

Exercise 7

7.1 thyroid, cricoid 7.2 epiglottis 7.3 thyrohyoid membrane
7.4 cricothyroid ligament 7.5 cricothyroid muscle

Exercise 8

8.1 vestibular and vocal folds 8.2 vestibular folds 8.3 vocal folds

Exercise 9

9.1 left primary bronchus and right primary bronchus
9.2 no; the right primary bronchus is wider and shorter
9.3 no; the right primary bronchus is more vertical and branches at a sharper angle than the left primary bronchus

Exercise 10

10.1 T 10.2 F 10.3 F 10.4 F 10.5 F
10.6 F 10.7 T

Exercise 11

11.1 visceral pleura
11.2 the visceral pleura dips into and lines the fissures of the lungs
11.3 the parietal pleura lines the ribs and vertebrae

CHAPTER 9

Exercise 1

1.1 T
1.2 F strike out lipase, replace with amylase
1.3 T
1.4 T
1.5 F strike out circular, replace with longitudinal; also strike out longitudinal, replace with circular
1.6 F strike out acetic acid, replace with hydrochloric acid
1.7 T
1.8 F strike out stomach, replace with duodenum
1.9 F strike out 12, replace with 20
1.10 F strike out jejunum, replace with duodenum
1.11 T

1.12 F strike out rugae, replace with villi
1.13 F strike out fat
1.14 F strike out parietal cell, replace with hepatocyte
1.15 T
1.16 T

Exercise 2

self-quizzing and pronunciations

Exercise 3

mouth, pharynx, esophagus, stomach, small intestine, large intestine, anus

Exercise 4

4.1 uvula 4.2 genioglossus 4.3 superior longitudinal
4.4 taste buds 4.5 esophagus

Exercise 5

drawing

Exercise 6

6.1 masseter 6.2 buccinator 6.3 facial nerve (CN VII)

Exercise 7

7.1 submandibular 7.2 sublingual 7.3 lingual and hypoglossal

Exercise 8

vagus nerve (CN X)

Exercise 9

9.1 contains large, longitudinal folds 9.2 rugae
9.3 to increase the surface area of the stomach

Exercise 10

10.1 large intestine 10.2 stomach 10.3 small intestine 10.4 stomach
10.5 small intestine 10.6 large intestine 10.7 stomach 10.8 large intestine
10.9 small intestine 10.10 large intestine 10.11 small intestine 10.12 large intestine
10.13 stomach 10.14 large intestine 10.15 stomach

Exercise 11

11.1 diaphragm 11.2 right 11.3 falciform 11.4 round ligament
11.5 inferior vena cava 11.6 portal vein
11.7 proper hepatic artery with its right and left hepatic artery branches
11.8 drawing

Exercise 12

12.1 the duodenum encircles the head of the pancreas 12.2 spleen
12.3 kidneys 12.4 superior mesenteric artery and vein

Exercise 13

13.1 attaches to the greater curvature of the stomach and covers the abdominal viscera anteriorly
13.2 attaches between the liver and the lesser curvature of the stomach
13.3 liver

Exercise 14

14.1 transverse mesocolon 14.2 mesentery

CHAPTER 10

Exercise 1

1.1 twelfth 1.2 medulla 1.3 nephron 1.4 blood capillaries
1.5 renal corpuscle 1.6 pedicels
1.7 blood plasma, glucose, amino acids, potassium, sodium chloride, and urea
1.8 filtrate 1.9 glucose, vitamins, amino acids, water, and ions
1.10 ADH 1.11 micturition 1.12 aldosterone
1.13 erythropoietin

Exercise 2

2.1 renal fascia 2.2 perirenal fat 2.3 left 2.4 liver
2.5 suprarenal (adrenal) gland 2.6 psoas major muscle

Exercise 3

3.1 renal veins 3.2 no; left renal vein 3.3 hilum 3.4 renal vein
3.5 ureters cross anterior to the iliac arteries and veins

Exercise 4

drawing

Exercise 5

5.1 b 5.2 c 5.3 a 5.4 d 5.5 a

Exercise 6

6.1 pubic symphysis
6.2 prostatic urethra, membranous urethra, spongy urethra
6.3 sphincter urethrae muscle

CHAPTER 11

Exercise 1

1.1 b	1.2 d	1.3 c	1.4 d	1.5 c	1.6 c
1.7 c	1.8 c	1.9 b	1.10 c	1.11 d	1.12 c
1.13 a	1.14 b	1.15 a			

Exercise 2

seminiferous tubules, epididymis, ductus (vas) deferens, ejaculatory duct, urethra

Exercise 3

3.1 c
3.2 d
3.3 a
3.4 e
3.5 c
3.6 a

Exercise 4

4.1 skin
4.2 skin and dartos muscle
4.3 superficial scrotal fascia
4.4 external spermatic fascia
4.5 cremaster muscle
4.6 internal spermatic fascia
4.7 tunica vaginalis

Exercise 5

5.1 ilioinguinal nerve
5.2 external abdominal oblique muscle
5.3 superficial inguinal ring
5.4 medial crus forms medial border of superficial inguinal ring;
lateral crus forms lateral border of superficial inguinal ring;
intercrural fibers form superior border of superficial inguinal ring
5.5 internal abdominal oblique muscle

Exercise 6

6.1 pampiniform plexus of veins, artery to ductus deferens, ductus deferens
6.2 testicular vein

Exercise 7

7.1 ampulla
7.2 a. posterior to bladder b. inferior to bladder
7.3 ejaculatory duct

Exercise 8

8.1 glans penis
8.2 a. fundiform ligament b. suspensory ligament
8.3 the deep dorsal vein lies in the midline; a deep dorsal artery and deep dorsal nerve lie lateral to the vein on both sides

Exercise 9

9.1 corpus spongiosum, corpus cavernosum
9.2 corpus spongiosum
9.3 bulbospongiosus muscle

Exercise 10

10.1 broad ligament

Exercise 11

11.1 ovarian ligament   11.2 drawing

Exercise 12

12.1 a. ovarian artery and vein   b. uterine artery and uterine venous plexus

Exercise 13

13.1 the fornix surrounds the inferior part of the cervix
13.2 sphincter urethra muscle
13.3 labia minora
13.4 clitoris

Exercise 14

14.1 T	14.2 T	14.3 T	14.4 F	14.5 F
14.6 T	14.7 F